AQUARIAL

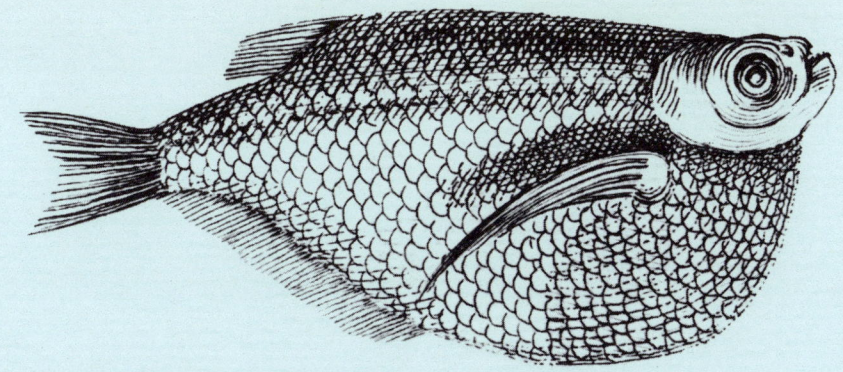

FISH

Introduction and Notes to the Plates
by Keith Banister, PhD,
Senior Scientific Officer, Department of Zoology (Fish Section),
British Museum (Natural History), London

Paintings by Tom Adams

FREDERICK MULLER LIMITED
LONDON

Cyprinus monstrosus
10

Edited by Ann Hill
Designed by Terry Smith

First published in Great Britain in 1979
by Frederick Muller Ltd, London NW2 6LE

Created and produced by Trewin
Copplestone Publishing Ltd, London
Design and text © Trewin Copplestone
Publishing Ltd, 1977

Filmsetting by Oliver Burridge
Filmsetting Ltd
Origination by Fotolithos, Zurich
Printed in Great Britain by Ben Johnson Ltd

ISBN 0 584 10369 7

Illustrations appearing on the preliminary pages of this book are as follows. Half-title page : a Three-spined stickleback by Thomas Bewick, from a chapbook published in the 1780s in Alnwick, Northumberland. Title page, from top left, reading clockwise : carp from volume V of Edward Donovan's "Natural History of British Fishes" published in London in 1808 ; a freshwater Flying fish from the "Zoophylacium" of Grenovius, published in Holland in 1763 ; an Albula from M. E. Bloch's "Systema Ichthyologica" published in 1801 with hand-coloured engravings by J. F. Hennig ; a spiny eel from Grenovius' "Zoophylacium" ; Poecilia vivipara from M. E. Bloch's "Systema Ichthyologica" ; a barbel from volume I V of Thomas Pennant's "British Zoology" published in Warrington in 1776. This page : Cyprinus monstrosus from H. Ruysch's "Theatrum Universale Omnium Animalium" published in Amsterdam in 1718. Opposite : marine monster in monk's clothing, an illustration from Guillaume Rondelet's "L'Histoire Entiere des Poissons" published in Lyon in 1558.

Contents

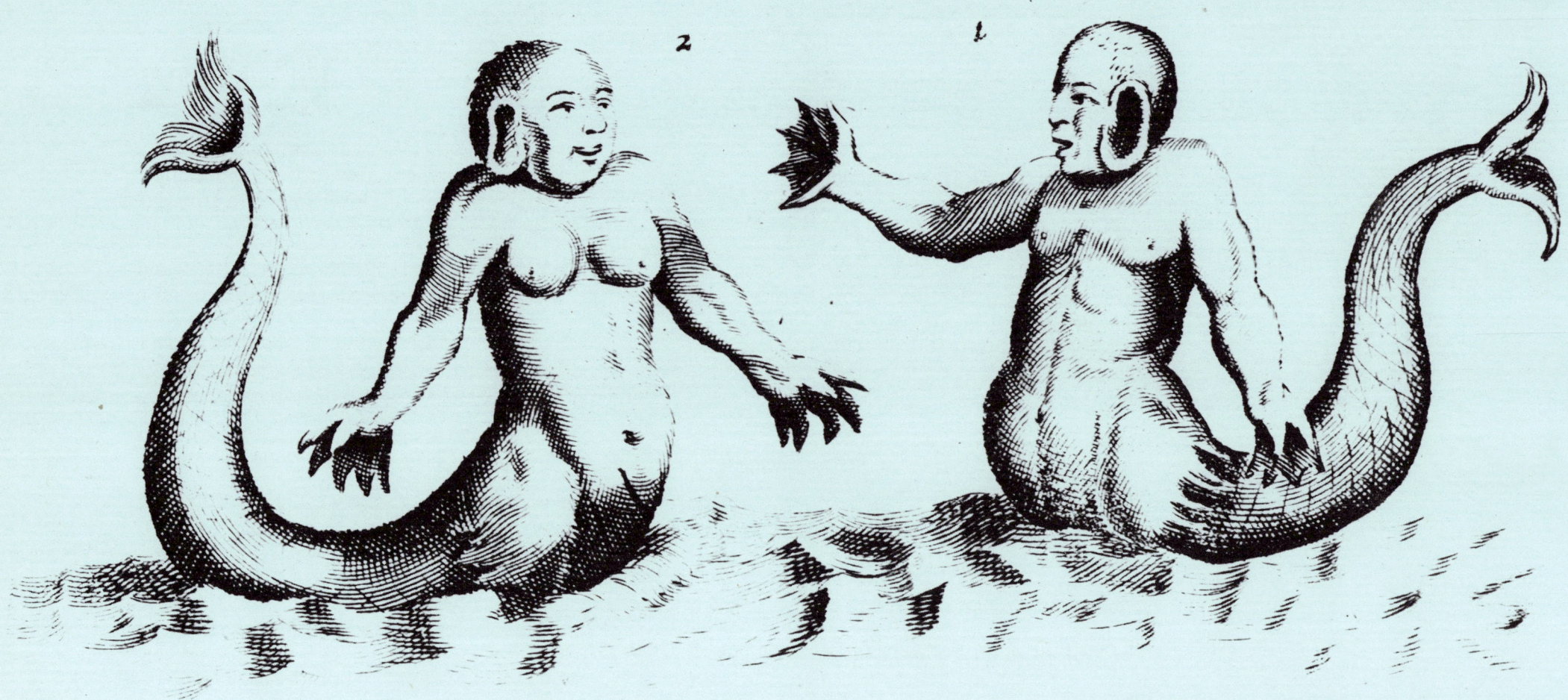

Anthro pomorphos

Anthropomorphos. One of the illustrations in H. Ruysch's "Theatrum Omnium Animalium" published in Amsterdam in 1718.

Introduction

"Perhaps the deities of the sea claim affinity with him, and carry him down into ocean depths; where, with the nobles of the sea and their attendant Nereids, he holds strange festivals in submarine palaces. Perhaps the Naiads of the river become enamoured of him, decoy him beneath the waves which they rule, and keep him, a very willing prisoner in their mysterious home, enchained by their beauty, and forgetful of his earthly love."

The Fresh and Salt Water Aquarium
Rev. J. G. Wood, 1867

Human beings have always had an interrelationship with animals. Initially, our early ancestors were an equal part of the fauna, sharing the vicissitudes of life with all the other animals in the environment. Early man hunted animals for his food, and equally some of the animals hunted man; the relationship was of a two-way, predator-prey type. As man's intelligence increased, the nature of his relationship with animals began to change. The various changes may have happened independently, more than once, in different parts of the world; we do not know, and probably never will. Nonetheless, from the few scraps of evidence available we can construct some pattern of the changes and the likely sequence of events.

His increasing intelligence enabled man to use the world around him to his advantage. Hunting is a time-consuming occupation; it is also dangerous. Both time and danger could be reduced if the main prey, the edible terrestrial mammals, could be to some extent confined in a known area where they could be bred and also cropped. Thus, the process of domestication started (domestication in this sense simply meaning changing the animal's way of life to suit man's needs). In all probability, the first animals so treated were herbivores. The advantages in this are obvious – their food grows naturally, they are inherently less dangerous than carnivores and they live in herds. If a group of early men, family unit or tribe moved from one site to another, then the herd could be taken along and food would be available, growing en route for the animals and on the hoof for the people. Goats, sheep and cattle were among the first animals used in this way. Man and dog have a relationship of long standing, but the original nature of the relationship is enigmatic. Although a few primitive tribes will eat dog, it is usually argued that man and dog formed an early working relationship to their mutual convenience. Gathering and protecting herds of sheep, goats or cattle, possibly with the assistance of dogs, resulted in the tribe or group of men developing a greater degree of independence from the environment.

The development of agriculture (although how this relates to either the timescale or geographical location of animal husbandry is not clear) presupposed the development of settled communities. Agricultural communities are not by definition continuously nomadic, although at first they may have been spasmodically so. With less time spent in travelling, more

Egyptian fishing scene. Fresco from Beni Hassan. Old Kingdom, 12th dynasty. (Photograph Mansell Collection)

[8]

time could be devoted to the nurturing and maintenance of the food supplies. Compounds could be built which would prevent the animals from straying and also deter the predators. Waste agricultural products could be used as food for the captive herds during winter or the dry season, again producing a certain independence from the environment. In a settled community, mankind confined and utilised, for his own advantage, that part of the world around him that he best understood.

Man is a terrestrial, not an aquatic being. Therefore all the domestication that he first practised was concerned solely with terrestrial animals. The watery environment of the aquatic animals was alien, and for a long time was used solely as a hunting-ground, not a farmyard. But with the increase in experience and technological ability this was to change – as a food source, the waters could not be neglected.

Fish formed an adventitious item of diet not only for man, but also for his ancestors. Most omnivores will eat fish if they have a chance. The Alaskan brown bear, for example (not in any way a claimant for hominid ancestry), has been seen to scoop up salmon from shallow rapids. The technique of catching fish in this way is sufficiently important for mother bears to teach it to their cubs.

In South America, the jaguar has been reliably reported to use its tail as a lure for fish. When the fish bites the tip of the tail, the jaguar flicks the fish onto the bank with its forepaw. The jaguar usually fishes below a fruit tree where the fish congregate to feed on the fruit falling into the water. Foxes, racoons and coyotes have been credited with similar fishing techniques. In the Trobriand Islands, rats have been seen luring crabs with their tails.

Early man is known to have been catching and eating fish at least 25,000 years ago. Fish hooks, belonging to the Magdalenian culture have been found, as have the remains of fish meals in the settlements. Arguments have been put forward that fish nets are as ancient in origin as fish hooks. Extremely ancient harpoons, beautifully carved in bone, ivory or stone have been found by archaeologists.

We do not know when, nor where, it first occurred to someone that confining fish in a small pond made it easier to catch them again when they were needed. Several natural phenomena may have initiated the idea, or suggested a parallel with the enclosure of sheep or goats. For example, during floods, fish can be washed into, or can swim into, pools, ponds and ox-bows, and then be contained there as the flood waters recede and the river falls to its normal level. During a dry season, the fish would have been observed to be concentrated in the small, isolated pools left in the otherwise dry river bed. Under these circumstances, not only are the fish easy to capture, but their ability to survive in small pockets of water would have been noticed. The step from observing such a natural phenomenon to artificially reproducing it for beneficial purposes is a small, but very important one. From the first muddy ponds, man's one-sided relationship with fish has grown into a multi-million-dollar concern.

Nowadays, the term "fish culture" includes several distinct aspects of the cultivation of fish, all of which have an economic angle. There is the ornamental aspect – the keeping of fish in aquaria for aesthetic purposes; the end result, however, relies on a chain of collectors, breeders and exporters all of them involved in procuring and producing fish for the pleasure of the aquarist. There is the sporting aspect of fish culture – the raising of fish in captivity for restocking overfished rivers and reservoirs. There is the food production aspect, extremely important in many parts of the world, which involves not merely breeding and rearing fish to an edible size in ponds, but can also involve raising young fishes through their most vulnerable stages before releasing them into the sea. There is the scientific aspect – keeping fish in captivity to discover how they react to their world and also to acquire pure knowledge of the way fish live and how their bodies work. Not all of these aspects are mutually exclusive. Indeed, there is a great deal of interdependence between them.

When fish were first held in captivity, the prime motive was doubtless to provide food. However, that would in no way have precluded the possibility of people deriving pleasure from the process of catching the fish or admiring their form and colour, or even observing their habits. But it was not until the development of more leisured societies that the non-alimentary aspects assumed importance in their own right.

The oldest known fish ponds were built by the Sumerians about 4,000 years ago. They were artificial ponds associated with temples, and there is no evidence that they functioned as anything other than a larder. We do not know for certain what species of fish were kept and the Sumerians did not appear to make any provision for breeding the fish. Fish ponds were also built by the Assyrians and it seems that these were a fairly common feature in larger settlements throughout the Fertile Crescent. There is some indirect evidence that,

A Red mullet. One of the plates in volume I of Edward Donovan's "Natural History of British Fishes" published in London in 1808.

THUS have I sung, how scaly Nations rove,
What Food they seek, what Pastures they ap-
prove;
How all the busy Wantons of the Seas
Soft Loves repeat, and form the new Increase.
But whence could Man the wond'rous Secret know? 5
To some kind Pow'r he must the Blessing owe,
Who to his View the hidden Depths expos'd,
Uncover'd all th' Abyss, and the vast Scene disclos'd.
For what great Work has Man unaided wrought?
Heav'n gives the Means, and Heav'n inspires the
Thought. 10
Did not assisting Influence from above
With unseen Force the passive Agents move,

K 2   The

A page from Oppian's "Haleuticks of the Nature of Fishes and Fishing of the Ancients" with an account of Oppian's life and writings and a catalogue of his fishes. This translation from the Greek was published in Oxford in 1722.

although they kept them, the Assyrians had no great interest in fish. Fish are figured on some of their seals, but whereas the boats that are also on the seals are drawn with care and accuracy, the fishes are stereotyped. They are so poorly drawn that it is impossible to be certain to which family they belong. One gains the impression that the fish are included merely to provide evidence that the boats are floating on water.

The Egyptians not only kept fish in ponds but, fortunately, painted detailed and accurate pictures of the fishes concerned. The paintings and reliefs on the walls of the tombs, which reminded the dead of the world they left behind, are precise enough for us to know that, for example, *Tilapia* were kept – a food fish still in considerable demand in North and East Africa. To the Egyptians, the fish was sacred; it was a symbol of fertility, and one of the forms of Isis is depicted with a fish on her head. Certain species of fish were taboo to the priests as well as to ordinary Egyptians. In the eighteenth dynasty, some fishes were embalmed. The Nile perch (*Lates niloticus*) was worshipped as a god and the town of Esueh was renamed Latopolis by the Greeks on account of the thousands of Nile perch embalmed there. Greek historians have recorded certain curious dietary customs of the Egyptians: fish were usually served with the fins removed; on the ninth day of the month of Thoth, the Egyptians ate fried fish (doubtless from their fish ponds) at the doors of their houses; the priests, however, offered their helpings to the gods by burning them. It is during this period that we have the first evidence of fish having a recreational purpose. Tomb paintings show young Egyptians fishing in ponds with a rod and line. This form of angling must have been for sport, for much more efficient ways of catching for food had already been developed. Nourishment and sport seem to be all that the Egyptians required of their living fish; in return, the dead fish were worshipped.

The Romans were well aware of both the beauty and utility of fishes, and their relatively advanced technology enabled them to exploit fishes more extensively than the Egyptians could. To us, living in the twentieth century, the uses that the Romans found for fish are considered bizarre, but at the same time we must admire their ability to utilise technology to the full.

According to Pliny the Elder (AD 23–79) keeping fish in *vivaria* was started before the end of the second century BC by Licinus Muraena (a rich man whose name is now commemorated in the scientific name for the Moray eel, *Muraena muraena*). Marcus Terentius Varro (116–27 BC) had two oblong piscinae in his aviary at Casinum, and in his book 'De Re Rustica' he distinguished two types of fish ponds: *dulces*, freshwater ponds maintained by ordinary people for profit – the fish were sold at market; and *salsae* or *maritimae*, seawater ponds maintained by the rich and noble and used solely as a spectacle. By the Late Republic and Early Empire, fishes were commonly kept in artificial enclosures for show.

Seawater ponds were also used to provide food as well as a cabaret – in some instances simultaneously. To the Romans, the joy of eating Red mullet (*Mullus surmuletus*) was greatly enhanced if the fish were brought alive to the table, so that the flickering iridescences and colour changes of the dying fish could be appreciated before the fish was eaten. A large, or beautifully coloured specimen could cost as much as a cow.

The "status symbol" fish for a Roman was the Moray eel, a large mottled species, common in the Mediterranean. Pliny the Elder reports that Gaius Hirrius so loved his Morays, which he kept in a special pond, that he declined to sell them to Julius Caesar (arguably a dangerous refusal). He did, however, lend them to the emperor and there is an account of 6,000 eels being on show at a sumptuous banquet. Wealthier Romans bedecked their pet Morays with jewels and the author Martial described eels responding to their masters' voices. Vedius Pollo (no doubt among others) entertained his guests by letting them watch his Morays being fed on refractory slaves.

The hedonistic examples of Roman fish-keeping described above are only a part of the picture; there is a practical side. The Romans experimented with serious piscicultural techniques by transferring the fertilised eggs of desirable or advantageous species to waters where they were needed.

Glass making was known to the Romans who produced many fine bottles and containers. If the dying Red mullet were first viewed alive in a glass container then this would rank as an early aquarium – albeit for a rather specialised purpose. However, there is no record, as far as is known, of the Romans maintaining live fish in glass vessels.

With the decline and ultimate collapse of the Roman Empire, the western world entered the dark ages. For about a thousand years there were very few books and in these books very few references to fish, let alone fish-keeping. In the West, it was not until the nineteenth century that "aquaria", in various guises, were as common as they had been during the Roman era. As the West declined, the East ascended and

Netting fishes and dog fish. Two woodcuts from "Tractatus dei Piscibus" in the "Hortus Sanitatis" edition of 1511, printed in Venice by Bernadinium Benalium.

A page from Dame Juliana Berners' "A Treatyse of Fysshynge wyth an Angle", part of The Book of St Albans printed by Wynkyn de Worde, Westminster, in 1496.

Salamon in his parablys sayth that a good spyryte makyth a flouryinge aege; that is a fayre aege & a longe. And syth it is soo: I aske this questyon. Whiche ben the meanes & the causes that enduce a man in to a mery spyryte.: Truly to my beste dyscrecōn it semeth good dysportes & honest gamys in whom a man Joyeth wythout ony repentannce after. Thenne folowyth it þ gode dysportes & honest games ben cause of mannys fayr aege & longe life. And therfore now woll I chose of foure good dysportes & honeste gamys, that is to wyte:of huntynge:hawkynge: fysshynge:& foulynge. The beste to my symple dyscrecōn whyche is fysshynge: callyd Anglynge wyth a rodde: and a lyne

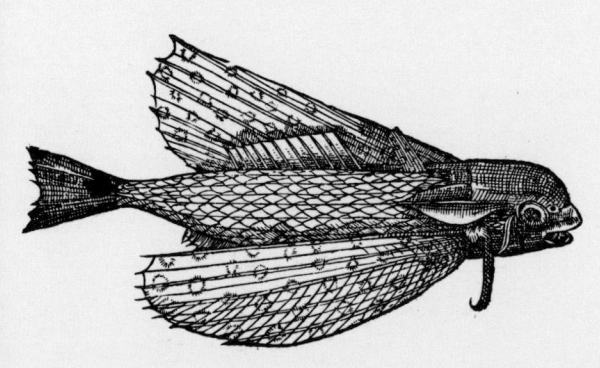

Flying fish, perch and carp from "L'Histoire Entiere des Poissons" by Guillaume Rondelet published in Lyon in 1558.

the Chinese began to cultivate and breed fish. However, the story of the development of the goldfish by the Chinese will come later and we now continue with the history of fish keeping in Europe.

Before the Norman conquest of Britain, the few references to fish mostly concern the Common carp (*Cyprinus carpio*), a freshwater fish of Europe and Asia, related to the goldfish, and one of the most important freshwater food fish. Towards the end of the Roman Empire, the natural historian Ausonius (AD 310–393) did not list carp from the Rhine and the Mosel rivers, although they are found there today. Cassiodorus (AD 490–585) described shipments of live carp from the Danube to Ravenna in Italy, for the table of the gothic king, Theodoric. The lessons of the Romans in the art of shipping fish had not at that time been completely forgotten. Charlemagne, king of the Franks, kept carp in ponds and sold them in markets. From these fragments of information, we can deduce that the carp is a fish that is easily transported alive, that there was a demand for fresh fish (especially in areas away from the coast), and that fish ponds were the prerogative of the rich and the clergy. In the Domesday Book we read that the abbot of St Edmunds had "II vivariae piscinae in villi ubi quiescat humatus S. Edmund" for the monastery dining-hall "ad victum monachorum".

Some rich non-clerics also had their own fish ponds. Again in the Domesday Book we learn that Robert Malet of Yorkshire had XX piscinae assessed at XX anguillae (eels).

A tract written about 1289, largely for the instruction of water bailiffs and gamekeepers says "every prudent man stocks his ponds, pools, lakes and reservoirs from his fisheries with Bream and Perch, but not with Pike, Tench or Eels, which devour an inordinate number of fish." Carp are not mentioned.

Geoffrey Chaucer (?1340–1400) wrote in the Franklin's tale:

"Full many a fat partridge he had in a mewe (cage)
And many a bream and many a luce in a stewe."

The "luce" was the pike (*Esox lucius*) and the "stewe" was the stew pond, a pond maintained, particularly in monasteries, for edible freshwater fish. The word stew is thought not to refer to the concentration of fish nor to the ultimate consumption of fish in a stew, but derives from the old French word estui, meaning to shut up or keep in reserve.

The stew pond was an important part of the monastery and

monastic life; the Church forbade the eating of meat on Fridays, so fish were eaten instead. Carp and tench are both tasty fish and well suited to life in small ponds; both were therefore eaten by the monks. Chaucer's Franklin, however, may not have had a good yield of fish from his pond if it included the piscivorous pike.

The clergy, and others rich enough to own fish ponds may well have played an important part in the distribution of the carp around Britain, for there is no certainty that it is a native fish. There is an old couplet:

"Hops and Turkey, Carps and Beer
Came into England all in a year"

from the chronicle of Sir Richard Baker; the year in question is presumed to be about 1530. The mention of turkey dates the couplet from after the discovery of America, because the turkey is an American bird. However, in 1496, Dame Juliana Berners, an abbess at St Albans wrote the first ever book devoted to angling, "a Treatyse of Fysshyinge wyth an Angle". Of the carp, the authoress wrote "the carpe is a deyntous fysshe: but there ben but fewe in Englande. And therefore I wryte the lesse of hym. He is an evyll fysshe to take." This statement implies that the carp was scarce and probably of limited distribution. If it had been introduced, then it was before 1496, but not all that long before.

There is another, probably contemporaneous, version of the couplet quoted above (but its source appears to be untraceable), that reads:

"Hops and Heresy, Carps and Beer
Came into England all in a year."

If the dating of the couplet (*c.* 1530) is correct, it would approximate to the time of the dissolution of the monasteries by Henry VIII. Carp coming into England could refer to the removal of the fish from the monastic stew ponds into the rivers and lakes, while "Heresy" needs no explanation. Certainly the carp was limited in its distribution in 1496, but widespread two hundred years later. It is to be hoped that archaeologists will find evidence in the mediaeval kitchen refuse heaps to solve this riddle.

One of the first fish books of the Renaissance was printed in 1554 and was written by Guillaume Rondelet, a professor of medicine at Montpellier university. In his writings, he

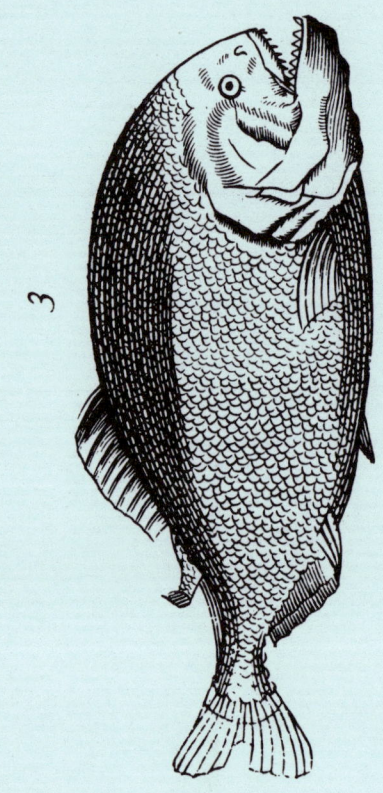

3

1. Camaripuguacu . Brafil .
2. Uubarana . Brasiliens .
3. Piraya et Piranha . Bras :

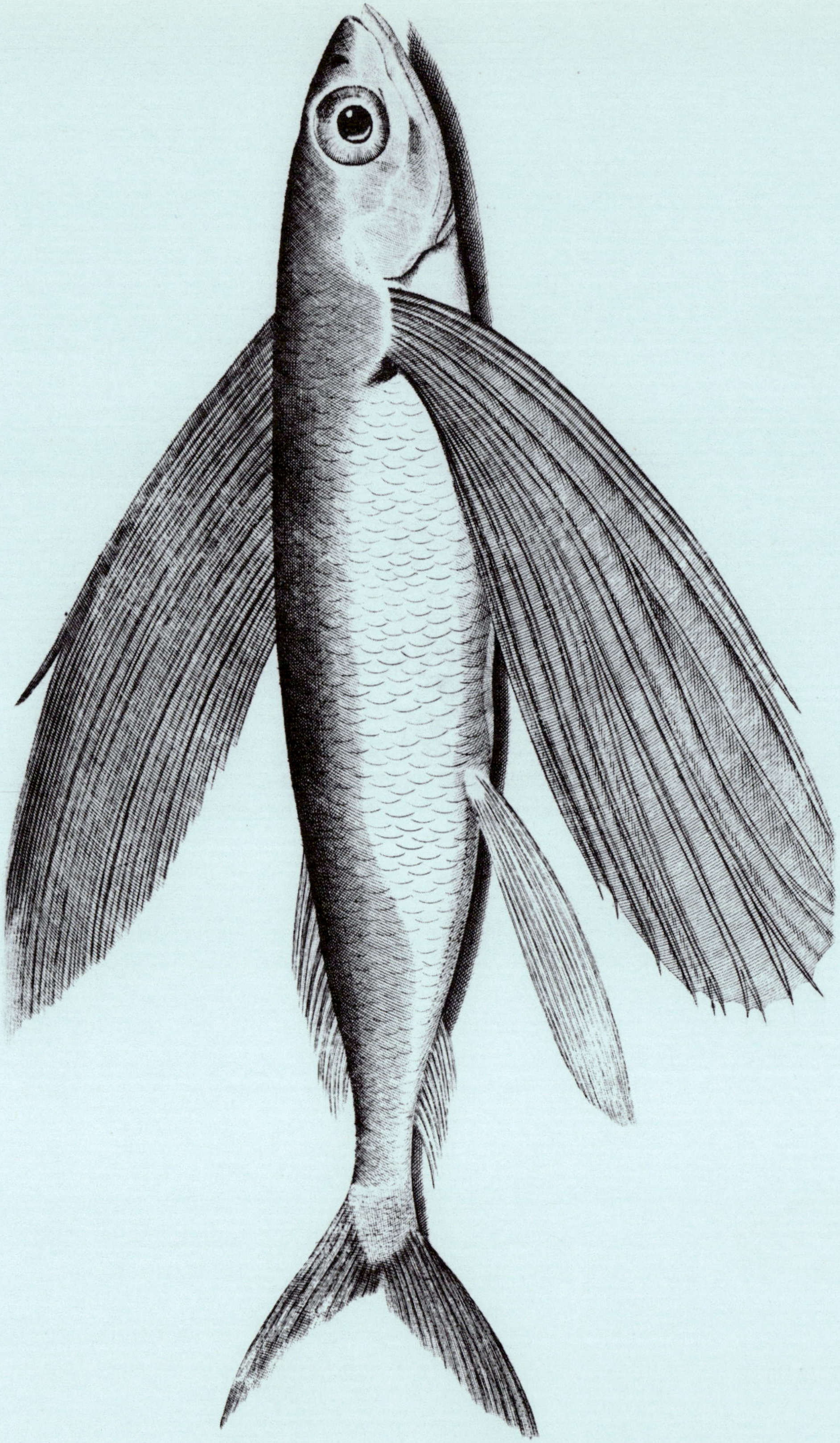

*A piranha and a marine Flying fish from
Francis Willughby's "De Historia Piscium"
published in Oxford in 1686 for the Royal
Society. Both of these plates were subscribed
by Samuel Pepys.*

Sumptibus D. Sam. Pepys Praes: S. R.

DIVISION I.
FISHES WITH A BONY SKELETON.

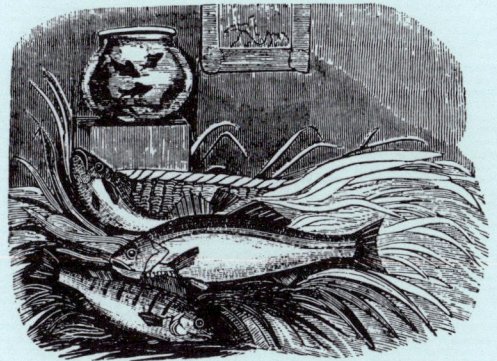

OF SPINOUS-FINNED FISHES.
(Order ACANTHOPTERYGII.)

One of the earliest illustrations of an interior to show a domestic glass bowl with fishes. It was reproduced in "The Book of Fishes" published in London in 1837 by the Society for the Propagation of Christian Knowledge.

Mr Warrington's tank. One of the plates in Arthur M. Edwards's "Life Beneath the Waters, or The Aquarium in America" published in New York in 1858.

stated that his wife had kept a fish alive in a glass container for three years. Although two other books on fishes appeared in the same year (by Pierre Belon and Hippolyte Salvian) they lack any comments that could be construed as referring to aquaria.

The next significant, but puzzling, comment comes from the entry for May 28th, 1665, in Samuel Pepys's diary:

> "Thence to see my Lady Pen, where my wife and I were shown a fine rarity : of fishes kept in a glass of water, that will live so for ever : and finely marked they are, being foreign."

It is usually thought that the fish were goldfish, but there is no evidence for this. The excellent ichthyological works of Willughby (1686) and Ray (1720) do not list the goldfish among British species (but see the text for plate 8). C. W. Coates of the New York Zoological Society has suggested that the fish may have been Paradise fish (*Macropodus opercularis*), native to South-East Asia. The arguments in favour of this suggestion are : the fish is "finely" marked ; it is obviously "foreign" (i.e. of alien shape and colour) ; it can stand low temperatures and it has an accessory breathing organ which enables it to survive in oxygen-deficient waters.

William Arderon, a remarkable observer of natural history living in East Anglia in the eighteenth century, kept fish in glass jars. He sent a communication on this topic to the Royal Society in 1746, and it was he who first noticed the voracity of the common stickleback.

Aquaria in the modern sense, i.e. a glass container holding a community of animals and plants in such concentrations as to reflect as nearly as possible the conditions in the wild, were developed in the nineteenth century. Probably no one individual was responsible, but various people experimented with aquaria and improved and refined each other's techniques.

One of the earliest reports of establishing an aquarium along modern lines was written by M. de Moulins of Bordeaux, in 1830. M. de Moulins found that by keeping plants in the water where his fish and molluscs were, the latter were the stronger and healthier for it. Such statements excited scientific curiosity and resulted in experiments designed to determine why the fish and molluscs were "the healthier for it". At the British Association meeting in Cambridge in 1833, Dr Danbery showed that submerged aquatic plants emitted oxygen and absorbed carbon dioxide under the influence of light. In 1842, Dr Ward published a work showing that animals and plants might be placed in air-tight containers, and that each might be so adjusted as to breathe in what the other breathed out. What is believed to be the first attempt to keep sea water constantly fresh by the presence of living seaweeds was successfully carried out by Mrs Anna Thynne, in 1846.

These early observations were brought together and rationalised by Mr R. Warrington in 1850. In March of that year, Mr Warrington read a paper before the Chemical Society of London describing his experiments and success with aquaria. His aquarium was a simple glass bowl of fresh water, in which were two goldfishes and some *Vallisneria* plants. Later, some snails were added to eat the algae which had grown on the sides of the bowl and the leaves of the plants. That simple, stable aquarium started the "great aquarium rush" of Victorian England.

The first public aquarium was erected in the Zoological Gardens in London in 1852 and the public were admitted in 1853. (Incidentally, the first recorded photograph of a living fish was taken in this aquarium.) A few months later, a small public aquarium was opened at the Surrey Zoological Gardens. A public aquarium opened in Dublin about the same time had a novel innovation. The curator, Dr Ball, a singularly ingenious man, used the enthusiasm of the public to aerate the tanks. He so constructed air-bellows that the visitors to the aquarium worked them with their hands, as a sort of amusement, in the intervals of strolling up and down to examine the contents of the tanks. Within a few years, major cities in Britain and in Europe had their own public aquaria. Not all of these limited their attractions just to water life. The Westminster Aquarium, for example, although it staged a "Piscicultural and Maritime Exhibition" in 1877, soon became a place of entertainment. The visitor could see caged animals, and the popular entertainers of the time would vie with young ladies of easy virtue for the custom of the patrons. Even the Brighton Aquarium, a most laudable and worthy institution which at this time possessed a tank with more than 110,000 gallons of sea water, could entertain its visitors with more than fish. It made a profit from shows, pantomimes and operatic concerts as well as giants, midgets, freaks and a first-class restaurant.

Although public aquaria were extremely popular, they did not adequately satisfy the Victorian public's enthusiasm for

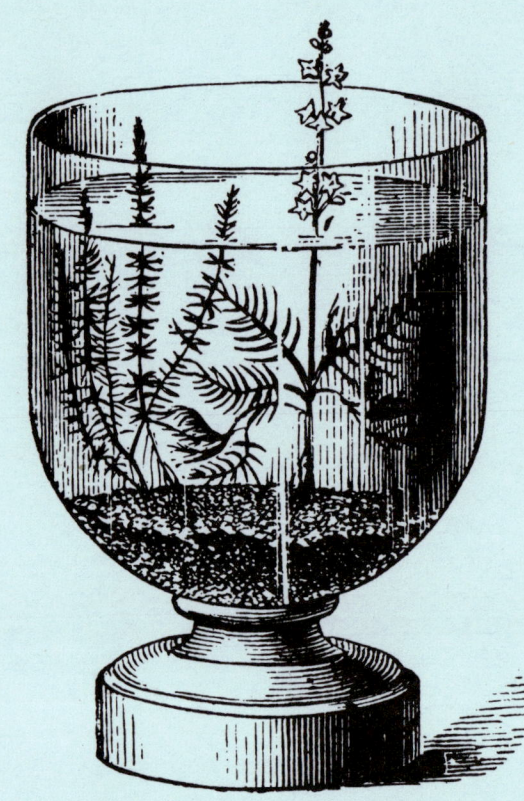

"Convalescent" Glass, containing *Hottonia*, &c. *Vallisneria* is the best restoring plant.

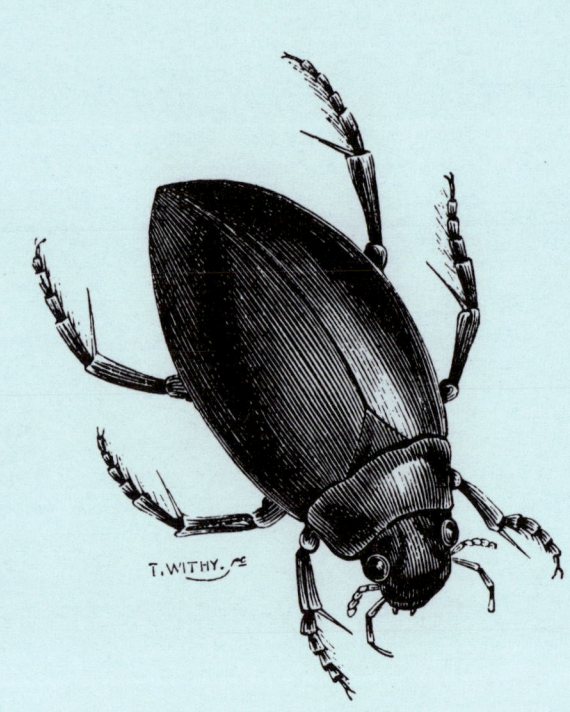

Great Aquatic Beetle (*Hydrophilus piceus*).

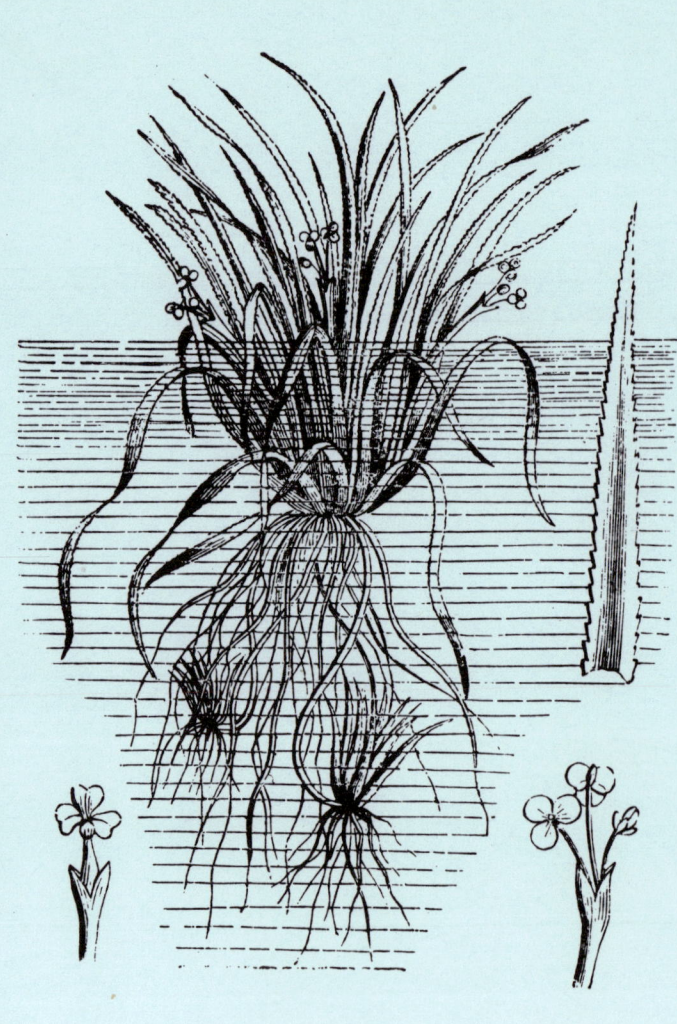

FIG. 18. CAN USED FOR THE TRANSMISSION OF FISHES, &C.

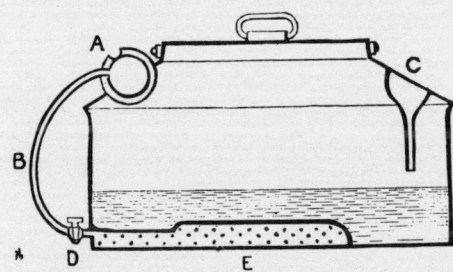

FIG. 19. SECTION OF CAN USED FOR THE TRANSMISSION OF FISHES, &C.

Mr Frank Buckland's apparatus for carrying fishes on a journey, illustrated in Reginald Bennett's "Marine Aquaria; Their Construction, Arrangement and Management, with full information as to the best animals and seaweeds to be kept, how and where to obtain them, and how to keep them in health" published in London in 1889.

aquaria; no home was complete without one and for a while the possession of an aquarium with interesting inmates was a status symbol. As a result of the public's desire for instruction, popular books on aquaria were written, the first three titles appearing in 1856. These were J. S. Hibberd's 'The Marine Aquarium'; Edwin Lankester's 'The Aquavivarium, fresh and marine'; F. S. Leach's 'Plain Instructions for the management of gold and other fish, water plants, insects etc.' and P. H. Gosse's 'The Aquarium or Unveiling the wonders of the deep'. America was two years behind. In 1858 two books appeared in New York: A. M. Edwards' 'Life beneath the waters, or the Aquarium in America' and H. D. Butler's 'The Family Aquarium'.

Peculiarly, these publications appeared almost exactly three hundred years after the first group of fish books (Rondelet 1554 and 1558, Belon 1554, Salvian 1554 and Gesner 1558). Any modern aquarist reading these books would consider them more as guides to the seashore than studies of aquaria. The very name "aquarium" was discussed and it was usually concluded that although the correct term "aquavivarium" should be employed, "aquarium" was more euphonious.

Almost any form of water life was kept. The aquarists in coastal towns would naturally concentrate on keeping sea fish, sea anemones, winkles, crabs and seaweeds. Readers of the early books on aquaria would be advised to feed sea anemones with chopped mussels, "to be offered in a pair of wooden tongs – much like sugar tongs." One is also advised that "the piked dogfish . . . is a variety so given to rendering itself an unsightly object by knocking its head against the boundaries of its tank, till it lays its cartilage bare, that it is frequently refused admittance to aquaria." As if it had any choice!

Although a great deal of the space in the early books is devoted to marine aquaria, there were more freshwater aquaria in operation than marine ones. The reason for this is that more people had access to freshwater animals and plants. The list of plants regarded as suitable for inclusion in aquaria reflects the lack of pollution in the rivers. The Water Soldier, for example, is now a rare plant, but in 1876 "is a very abundant plant in Norfolk and Suffolk as the anglers thereabouts have long since discovered." An interesting observation from the same source is that the Canadian Pond weed had been introduced but was not known to flower and *Vallisneria spiralis*, described as native to southern Europe

had been introduced to oxygenate the water in which medicinal leeches were kept and was by then widespread.

Water beetles were thought to be "most instructive" inhabitants of an aquarium, but J. E. Taylor added a note of warning about the giant water beetle that ". . . in the summer months . . . if the top of the aquarium be not closed, its owner may be somewhat astonished at seeing some of its aquatic inhabitants leaving it to dash themselves against the gasglobes!".

"Of course, it adds greatly to the interest of cabinet or other aquaria if the various specimens can be obtained personally from their native waters" wrote the Rev. Gregory Bateman in 1890. To these ends, the reader is advised to obtain a can with a mechanical contrivance for aerating the water within. A pleasant indication of the nature of Victorian society is given by the last-cited author who, while debating the merits of various types of carrying can wrote "the former is a German invention and although Mr Buckland [Frank Buckland] while speaking highly of this can, says that he engaged a tinman to construct it, he omits to mention the workman's name and address".

In the middle of the last century, aquaria were usually made, not bought. There may well have been firms who catered for the need to possess aquaria, but judging from the detailed building instructions given in the books available at the time, the suppliers were very few. Desirable aquaria of the period bore little relation to the oblong open-topped boxes of today, as the accompanying illustrations show. The aquaria could be pieces of furniture or artistic creations in their own right rather than devices for maintaining fish.

The simplest tanks of the period were little more than high-sided trays, with a base and three sides of wood, the fourth side of plate glass, the seasoned wood was screwed together and the glass attached by beading. Pitch was used to render the joints and the wood watertight. Under the direction of a plumber or carpenter, we are informed that such a structure can be constructed for less than fifteen shillings. A costlier method, but presumably longer lasting, was to substitute slate for wood. The "window aquarium" approximated to a modern aquarium but the maker was still advised to make only the front of glass. Even then "It should not be placed in a window having a south front, as then the light is too strong, and will develop that pest of the aquaria, a thick opaque green-ness, do what we may." Various bowls or "convalescent glasses, bell-glasses and stock-glasses" could be bought but

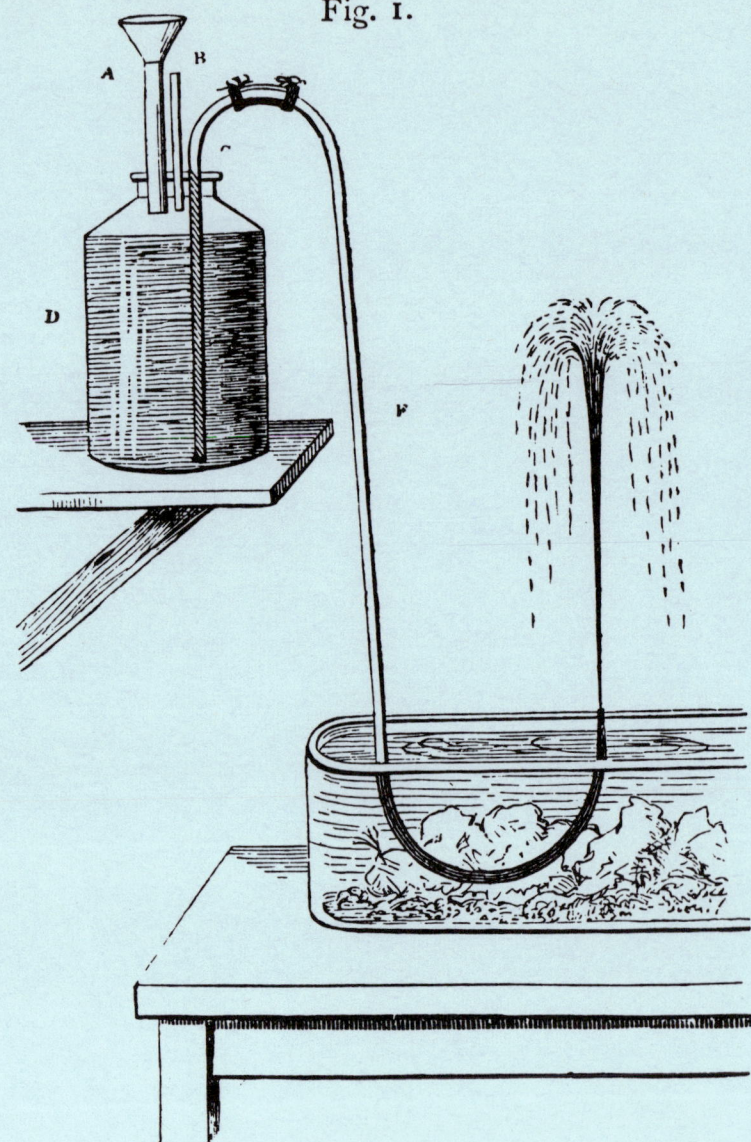

Fig. I.

An "extemporised fountain for small fresh-water or marine aquaria". From J. E. Taylor's "The Aquarium; its Inhabitants, Structure and Management".

Aquarium containing an "easily arranged fountain". Note the container on top of the bookcase and the jug concealed behind the curtain. From the Rev. Gregory C. Bateman's "Fresh-Water Aquaria: their Construction, Arrangement and Management, with full information as to the best water-plants and live stock to be kept, how and where to obtain them, and how to keep them in health" published in London 1889.

FIG. 37. CORRECT METHOD OF FILLING AN AQUARIUM.

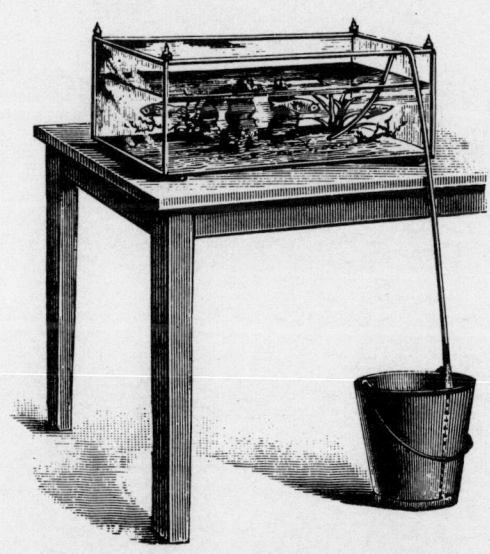

FIG. 38. METHOD OF EMPTYING AN AQUARIUM.

More practical advice from the Rev. Gregory C. Bateman in his "Fresh-Water Aquaria".

the oblong aquaria were usually home made to fit a recess. A resultant problem of the home manufacture of oblong "window aquaria" was that of making the joints watertight. "Leakage is a source of annoyance and untidiness in a room, besides interfering with that balance of animal and vegetable life which is sought to be sustained in water." The order of priority is quite revealing! Various methods of rendering joints watertight are advocated. Apart from pitch, one can make a cement of "fine white sand, one part litharge, one part resin, one-third part mixed into a paste with a little linseed oil and applied *unstintingly*." Another recipe is "one part, by weight, of rough *gutta-percha* and two parts of common pitch." Alternatively, if the frame of the tank were made of zinc, one could seal the joints with Aspinall's Bath Enamel.

Aeration of "streamless" aquaria was encouraged and widely practised. Pumps, as we now know them, had not been invented, but extemporised designs were given in many books. One such (from Taylor, 1876) is shown opposite. The instructions for the construction and operation are:

"All that is required is a wide-mouth bottle, in the cork of which are three holes through which the glass tubes seen in the sketch pass. C reaches nearly to the bottom, whilst the other two pass only through the cork. A is a wide, funnel-topped tube. C is bent at the top, and has there attached to it a piece of long indiarubber tubing. The cork and tubes should fit perfectly. In order to set this easily improvised fountain in action you fill the bottle. When it is full, continue pouring water gently into the funnel until it is above the level of the bend in the tube C. A little will then flow over into the long leg of the syphon E. The water will of course continue to flow until the level of the water in the bottle falls below the mouth of C. The tube B is for the escape of air whilst filling. A very short experience will enable the student to work this cheap fountain, and it is evident that it will flow for a greater or less space of time according to the magnitude of the water bottle and the bore of the indiarubber pipe, which is bent upwards at the extremity for the purpose of throwing the water into the air. It is true, more water is wasted by evaporation in this manner, but this is a difficulty easily met, as sufficient fresh water can always be put in the service or feeding bottle."

Generally, the recommendations were that the set-up aquarium should be as natural as possible but, by 1890, unnatural aquarium decorations were on sale. In a book published in that year, we read:

"this rockwork will also provide shade for the fish and the other animals which delight in retirement. It can be bought all ready for the aquarium, but often that which is offered for sale is gaudily coloured or in other respects unsuitable. Sometimes it will be found in the shape of ruins of various kinds. Of course such rockwork is altogether out of place in a well-arranged tank, for no one with any taste at all would care to see a fish, for instance, swimming through the window of a house, or a triton wriggling through the loophole of a castle."

What of the fish? In the early days of the aquarium boom indigenous freshwater fishes, invertebrates and plants were used (or marine forms if one lived near enough to the coast). The goldfish was the only species known to have been imported; there were, as yet, no tropical fishes. Nonetheless, there are some interesting references to hot-water fishes in some of the books. In an American book of 1858:

"the gold-fish, however, from being natives of warm climates, thrive best in water that would be uncomfortably warm to other species. Many of those exposed for sale are reared in water which receives the waste stream from factories, and is often kept as high as eighty degrees of Fahrenheit."

In England in 1890, the Rev. Bateman wrote:

"A great many Golden Carp (Goldfish) are now annually bred in this country, especially in those tanks which receive the waste hot water from some of those great manufactories in the North of England. For the aquarium 'cold-water' fishes should always be chosen – that is those which have been born in ponds uninfluenced by artificial heat: they are much hardier than the hot-water ones. Not seldom the latter will be found on their backs in the tank gasping for breath and apparently dying. When this is the case, they may be revived by placing them in running

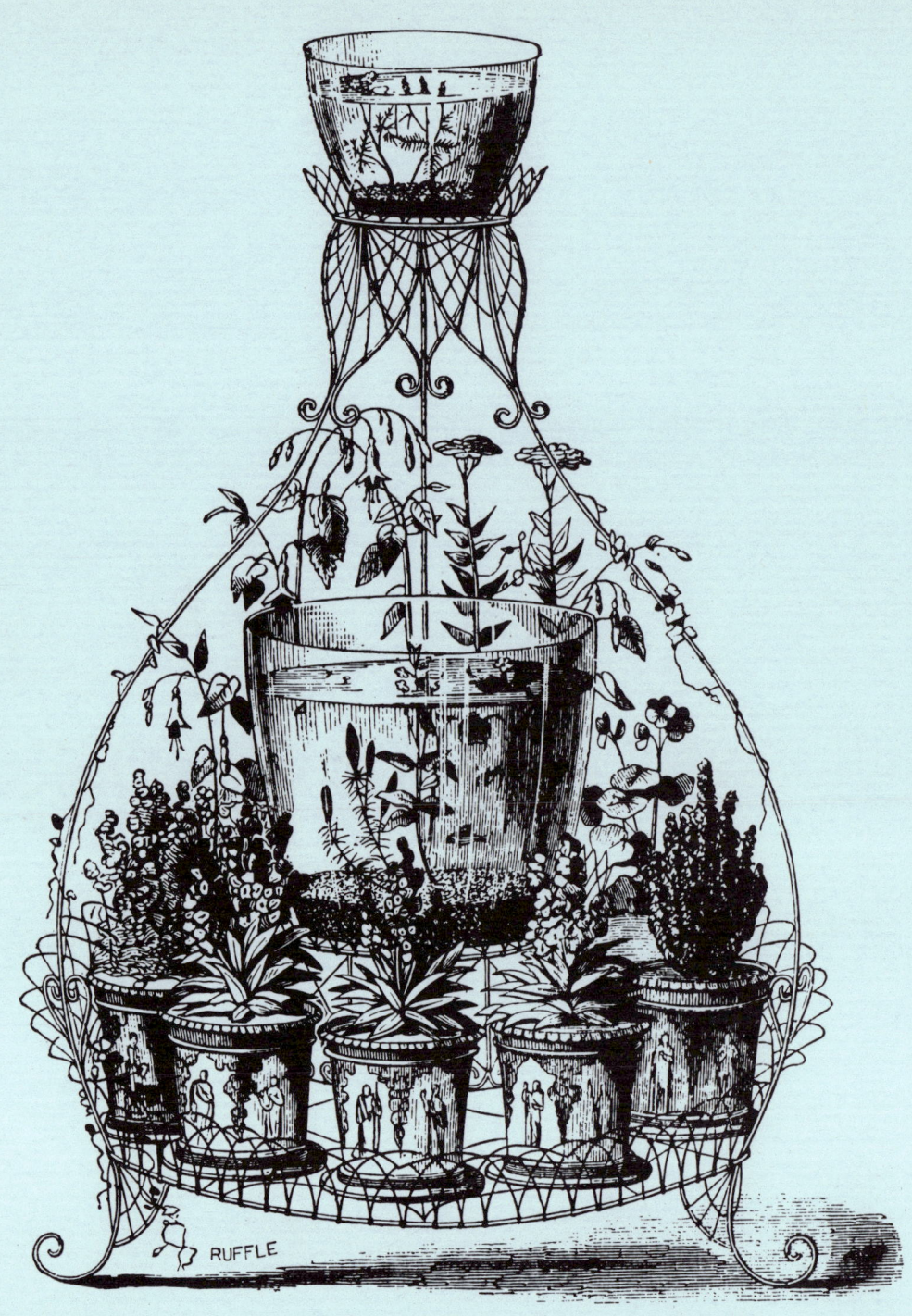

RUFFLE

Wrought iron flower stand with two bell glass aquaria, surrounded by potted plants.
From J. E. Taylor's "The Aquarium; its Inhabitants, Structure and Management".

An elaborate set-piece for a fashionable drawing room. From the Rev. Bateman's
"Fresh-Water Aquaria".

water, under a tap for instance, or by dropping a little brandy down their throats, or by putting them for sometime in fairly warm water (about 90 degrees). This last remedy – the simplest – is perhaps the most effectual.''

It is possible that the absence of unusual or spectacular fishes for the home aquarium contributed to the decline in aquarium keeping that started in the late 1850s.

The Rev. J. G. Wood (whose purple passage on the aquarium enthusiast is quoted at the beginning of this Introduction) wrote in 1867, describing the rise and fall of the recent mania:

"The fashionable lady had magnificent plate-glass aquaria in her drawing-room, and the schoolboy managed to keep an aquarium of lesser pretentions in his study. The odd corners of newspapers were filled with notes on aquaria, and a multitude of shops were opened for the simple purpose of supplying aquaria and their contents. The feeling, however, was like a hothouse plant, very luxuriant under artificial conditions but failing when deprived of external assistance.

Perhaps the beautiful plate-glass aquarium fell to pieces, discharged several gallons of sea-water over the fashionable carpets, and covered the fashionable furniture with sea-anemones, crabs, prawns and other inhabitants of the waters. Or some of the inmates died, and the owner was too careless to remove them. Consequently they were left in the holes into which they had retreated, and in a few days they avenged themselves for the neglect by rendering the water so fetid that no-one with ordinary sensibility could remain in the room. The schoolboy was very careful of his aquarium for a time, but in a month or two became tired of the constant attention required for its maintenance and so gave it up.

So, in due course of time, nine out of every ten aquaria were abandoned; many of the shops were given up, because there was no longer any custom; and to all appearances the aquarium fever had run its course, never again to appear, like hundreds of similar epidemics."

How wrong he was!

The interest in aquaria was revitalised with the introduction of exotic fish. Precisely when the first "tropical" fishes reached London is uncertain. The well-known writer of books on aquarium fish, Gunther Sterbar, quotes 1876 as the year that the Paradise fish was introduced into Germany, but this author has seen some colour plates, from an as yet unidentifiable German book, dated 1862, showing Paradise fish with live colours. The Rev. Bateman wrote in 1890:

"there are some very beautiful little fish called Paradise Fish . . ., which are in every way suitable for the aquarium. They were taken from a brook near Canton, China, and brought to France by M. Simon, who was the French consul at Ningpo. These fish, up to the time of their introduction into Europe in 1869, were said to be unknown to naturalists. M. Simon placed them in the charge of M. Carbonnier, under whose care they soon began to spawn. After the spawning, the male fish ejected bubbles from his mouth which did not break when on the surface of the water. He then, much to M. Carbonnier's astonishment, began to swallow the eggs, which he afterwards discharged under the shelter of the bubbles. The female fish was then driven by the male from the neighbourhood of the bubbles and ova, over which the latter fish kept a vigilant watch. In about three days the young fish were hatched, and soon began to feed upon cyclops, water fleas and the like. The Paradise Fish will breed more than one a year if the water is kept of a suitable temperature.''

Mr Frank Buckland, the eminent Victorian naturalist and an Inspector of Fisheries, possessed a pair of these fish and declared them the most beautiful fish he ever saw. Buckland was a member of the Acclimatization Society, a group of naturalists dedicated to introducing advantageous species to areas in which they were not known. Of the Paradise fish, he said:

"they would, I feel convinced, do exceedingly well in this country, not, of course, as fish to turn out but as aquarium – or to coin a phrase – hot-house fish. Should any lady take a fancy to looking after these

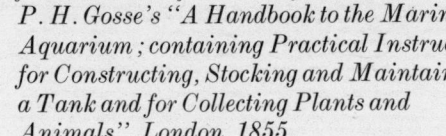

A handsome early Victorian aquarium, complete with ornamental rock-work and fountain. This formed the frontispiece to P. H. Gosse's "A Handbook to the Marine Aquarium; containing Practical Instruction for Constructing, Stocking and Maintaining a Tank and for Collecting Plants and Animals". London, 1855.

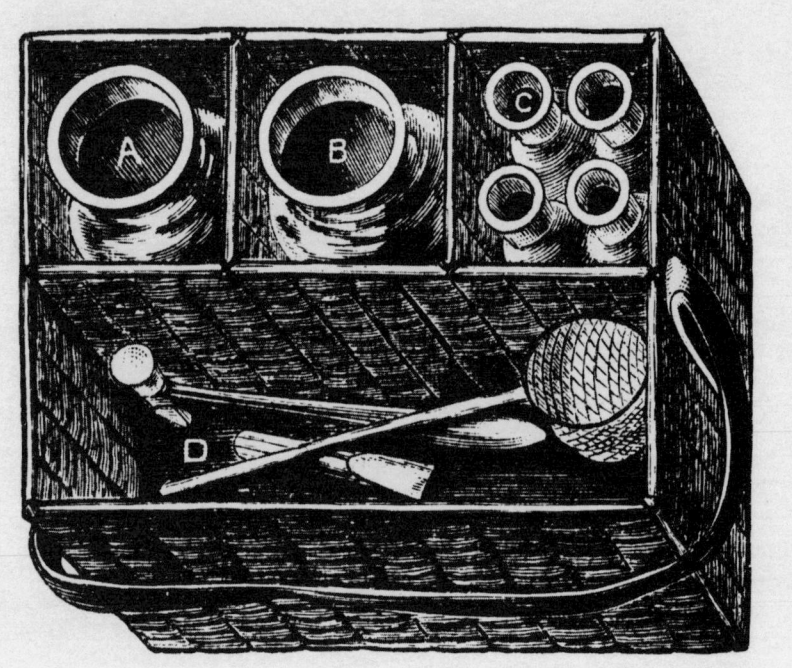

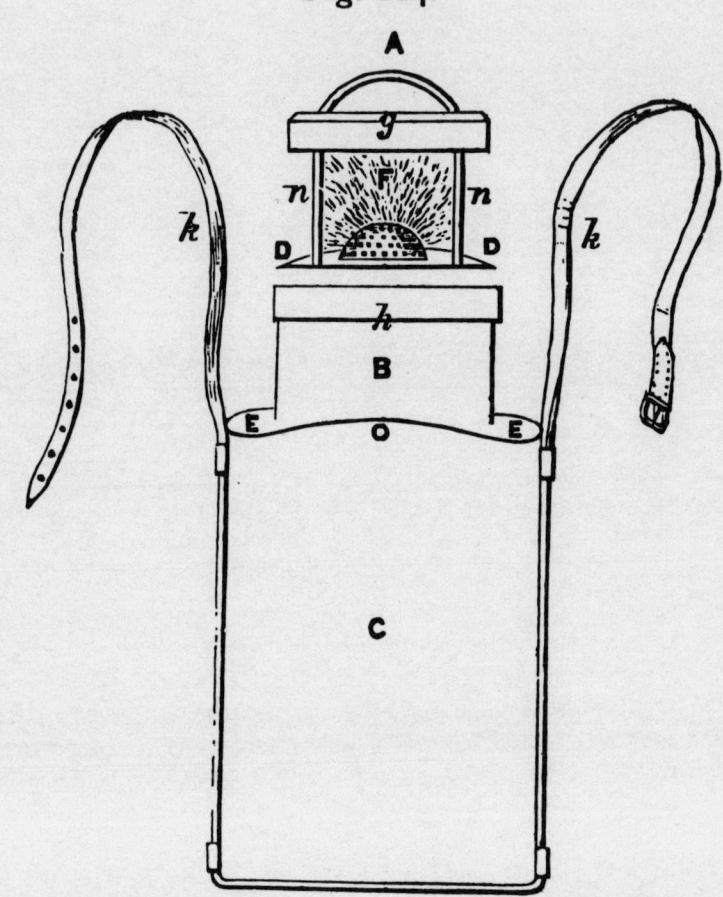

Self-acting Air-can for Aquatic Animals.

Fig. 115.

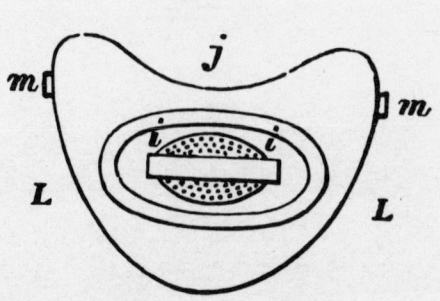

Section of ditto.

Design for a collecting basket for use when "shore hunting". From Reginald Bennett's "Marine Aquaria: their Construction and Management", 1889.

Mr A. J. R. Slater's self-acting aicran. "One of the best contrivances", says Mr Taylor, "that we have hitherto seen for keeping aquatic objects".

[22]

fish – keeping them in a warm room, feeding them regularly, and seeing the water is properly managed – they would, I am convinced, become acclimatised in this country. It has been proved that they will breed in France – why not, therefore, in England?"

The description of the Paradise fish bred by M. Carbonnier and kept by Frank Buckland is interesting – and very confusing:

"The scales glisten with all the colours of the rainbow. Their fins are most extraordinary: the dorsal fin extends from near the back of the head quite to the caudal fin; the anal fin is as long as the dorsal. Both fins have bony rays, which gradually increase in height from the commencement of the fin, the last ray in each fin extending to the extremities of the caudal fin. The caudal fin is large and forked, and also consists of bony rays; these rays have sharp points. The ventral fins each have a few short bony rays and one exceedingly long and apparently strong ray. The pectoral fins have soft rays."

The figure accompanying the description (reproduced here) shows a fish that lacks a forked tail and generally resembles a gourami of a genus such as *Colisa*. Paradise fish have an elongated ray in the ventral fin, but not to the same extent as do the gouramis. Perhaps one day an aquarist with a penchant for historical detection will be able to confirm whether or not Frank Buckland really did have Paradise fish.

In the books of the 1890s there is no mention of electric heaters and thermostats. Fishes to be kept warm were to be placed in a warm room. The date of the invention of the aquarium heater and thermostat is not known, but both were in use by 1913; so it was possibly about the beginning of the twentieth century that warm-water aquaria came into use with the public. With the development of techniques for maintaining easily a constant water temperature, the local freshwater fishes lost, to some extent, their appeal. Exotic fishes were wanted, the more colourful and bizarre, the better. Tropical plants and snails were also imported, not only to lend realism to the new environment of the fish, but also because they were adapted to the conditions.

Before the First World War, the only tropical fishes imported were those species that could stand the necessarily

long sea voyage. So the hardy species, e.g. Harlequin fishes and mollies, were amongst the first to be known in Europe. The actual dates of the first known imports are quoted elsewhere in this book. Two inventions have increased the selection of fishes available to European and American aquarists. These two inventions are jet aeroplanes and plastic bags. Formerly fish were transported in cumbersome, heavily-insulated cans. These were heavy, difficult to manage and, what is more, resulted in a substantial proportion of the fish arriving damaged from contact with the unyielding metal sides of the can. Nowadays, fish are densely packed in plastic bags which will not cause physical damage to the fish. More fish can be packed in one bag because the enclosed air is oxygen-enriched. The fish can even be given a sedative to reduce their activity and rate of oxygen consumption. The plastic bags are packed in ultra-lightweight polystyrene boxes to insulate and protect them so that the fish form a much higher proportion of the total cargo weight than hitherto. The speed of travel is now such that no importing centre is more than twenty-four hours away from the exporting centre. Fish losses are thereby minimised, serving to reduce the price of the fish in the pet shop.

The search for the new and the exotic continues. Remote areas are searched by the professional fish collectors and it is a far from rare event to find species unknown to science in the exporters' holding tanks. A sad corollary, from the scientific point of view, is that the original locality of the specimens cannot always be established.

As well as the demand for new or rarely imported species, there is also a desire for new varieties of well-known species. To this end, there are breeding establishments in many countries where strains of fish are carefully selected to produce a new colour variety, or a long-finned variety. Recent advances, e.g. the black Angel fish, and the long-finned platy, amongst others, are shown in the plates in this book and are mentioned in the text. Some breeders, on the other hand, just continue to supply the demand for the commoner species, so many of which are now bred in captivity that wild-caught specimens are specially identified in the shops and may command a higher price. There are some instances in which the fish breeders have, unwittingly, disturbed the local fauna. Hong Kong is a large export and breeding centre and many of the fishes now found in the streams of that island were originally from India, Malaysia or Africa. The fishes have escaped from the fish farms and colonised their new

A Paradish fish illustrated in the Rev. Gregory C. Bateman's "Fresh-Water Aquaria".

Leaping Carp

Yang Wei-ts'ung (from Hai-yen, Chekiang province). Ch'ing Dynasty, mid 17th century AD. Hanging scroll, ink on paper. Courtesy Trustees of The British Museum.

This impressive painting was undoubtedly painted as a chung-t'ang, *a large scroll to be hung centrally in the main reception hall of a dwelling, where it would face visitors as they came in. The subject is the Lung Men, or Dragon Gate, with a leaping carp symbolising the examination candidate who by his efforts will become an official, changing from fish into dragon. At court the officials wore dragon robes. It appears to be the only surviving example of Yang Wei-ts'ung's painting, although contemporary accounts tell us that his paintings were famous and sought after. The skill of the artist is evident in the careful control over ink and very faint colour washes in the stylised treatment of the subject.*

environment.

Such, then, is the current state of the tropical fish hobby. But before leaving the subject, it is necessary to back-track over 4,000 years to China, the country which has had the longest continuous history of cultivating fish for aesthetic purposes. The complete history is not known, but the available snippets are fascinating.

According to a recent report by a Chinese fishery worker, the artificial hatching of fish eggs was started about 2000 BC. The information is contained in a book by Fan Lai, dating from about 475 BC, on the culture of the common carp, and there is a suggestion that fish culture was aided by silkworm culture. Lai dates the origin of silkworm culture as 2698 BC and points out that the faeces and pupae of the silkworm are important supplementary items of diet for fish. The fish ponds were located below the silkworm colonies. If this information is correct, then fish culture in the East considerably antedates fish culture elsewhere. There are records of the Chinese awareness of red-coloured goldfish from about 350 AD. During the T'ang Dynasty (618–709) golden, fish-shaped badges were a sign of high office. Some of the other early references to goldfish are mentioned briefly in the text to plate 8. Goldfish were well known during the Sung Dynasty (960–1279) and were kept in captivity in houses from about the eleventh century onwards. There is a passage written by a monk in the tenth century which indicates a certain amount of elementary medical care of fish was known ". . . if they (goldfish) have poplar bark, they do not breed lice." Regrettably many of the early references are couched in obscure phraseology and this, coupled with the difficulty in translation, produces an uncertain, incomplete and confusing history. But no one can deny the fascination of the search.

Goldfish were brought to London in 1691 (*vide* Thomas Pennant, 'British Zoology', 1776), or at least that is the first recorded importation. In 1728, Sir Mathew Dekker brought a "great number" over and circulated them around the neighbourhood of London whence they were distributed to most parts of the country. Pennant also mentions that in China

"every person of fashion keeps them for amusement, either in porcellane vessels, or in the small basons that decorate the courts of the Chinese houses. The beauty of their colors, and their lively motions, give them great entertainment, especially to the ladies, whose pleasures, by reason of the cruel policy of that country, are extremely limited."

Goldfish were introduced into America during the eighteenth century, they very quickly spread over New England and now live wild in many of the states. The adaptability of goldfish is manifest in the ability of this species to live in waters of a wide range of temperatures, hence it lives in many countries. The introduction has usually been accidental, rather than deliberate; but deliberate introductions, to provide food, have been made. Even in its ancestral home, the goldfish was valued as food as well as an object of beauty. The history of the domestication of the goldfish is a long and complicated one, which can only be touched on here. However the reader who is particularly interested in the many, varied aspects of goldfish and their history, can do no better than read 'The Goldfish' by George Harvey and Jack Hems.

Unlike the rapid rise and fall of the aquarium mania in the middle of the last century, the current enthusiasm is far more likely to last. This is not merely because our advanced technology and greater knowledge enables us to achieve a greater success in keeping the fish alive, but also because there are so many different reasons for keeping the fish in captivity. Whether these reasons be scientific, economic or social, one overriding common factor is that the fish fascinate the watcher.

It is this fascination that has prompted Tom Adams to paint the following studies of fish and to capture so dramatically that graceful beauty which has attracted mankind for over four thousand years.

KEITH BANISTER,
June, 1977

The twenty-five colour plates in this book illustrate fifty-two different species. For compositional reasons, the fish are not painted to a common scale. On the ensuing six pages, they are all shown life size. The pen and ink drawings are by David Mallott.

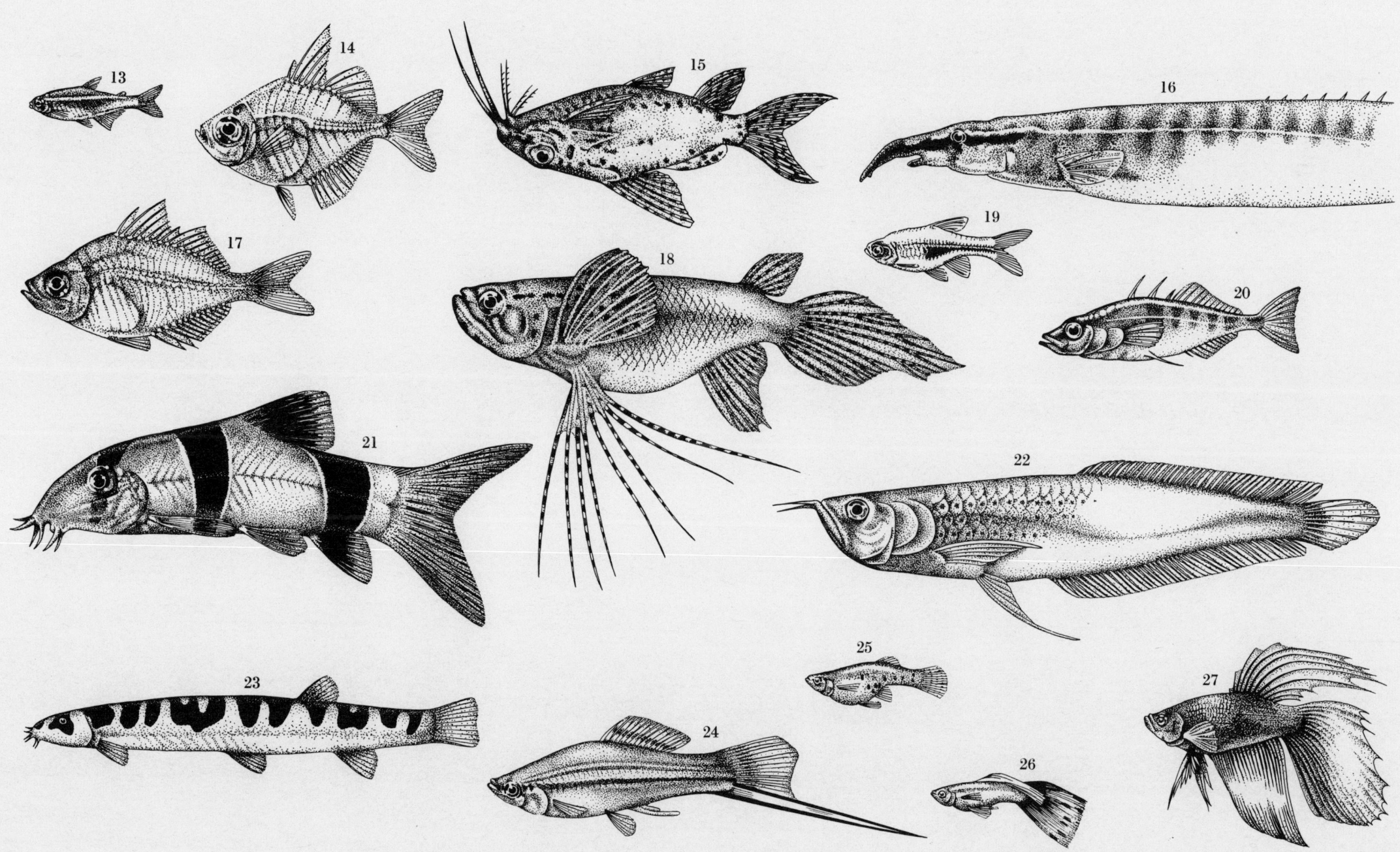

40

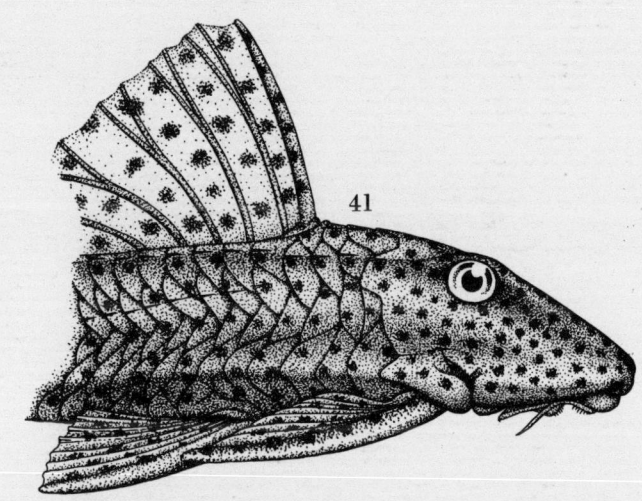

41

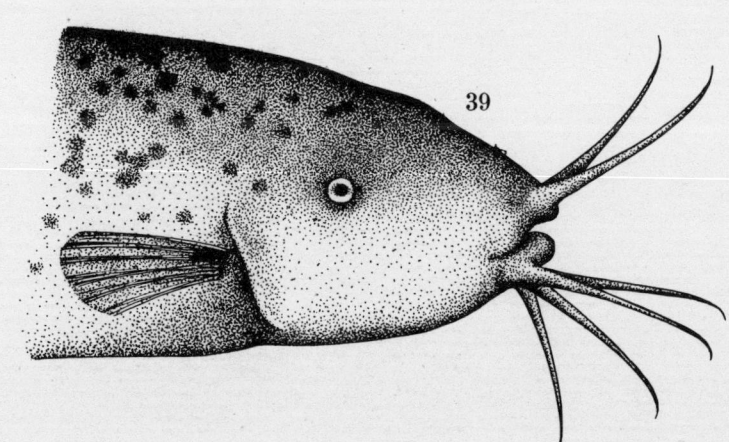

39

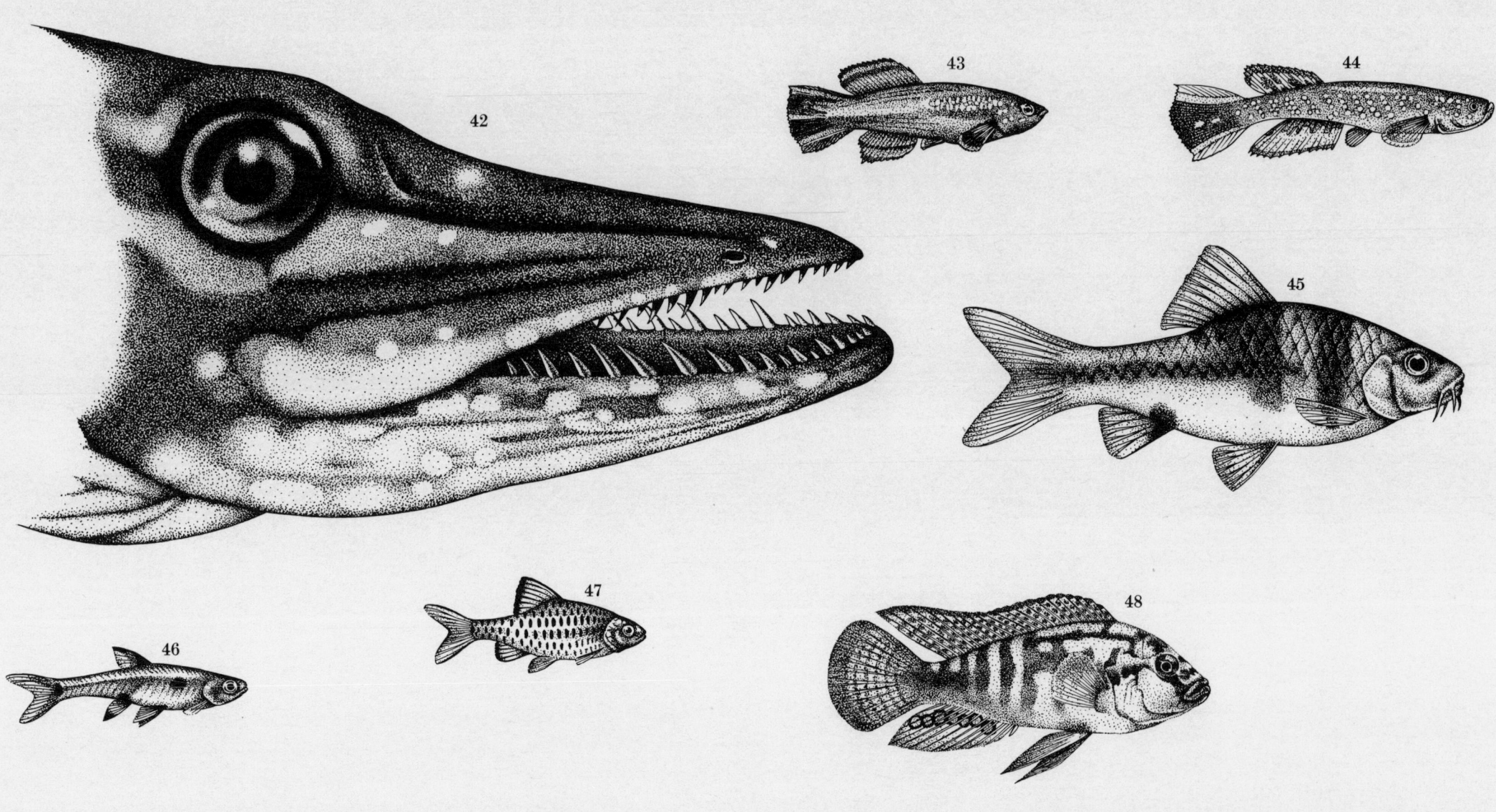

Key to the line drawings on the previous six pages

The Plates

Paintings by Tom Adams

From the rivers of Europe. Perch (Perca fluviatilis) *below left,*
pike (Esox lucius) *centre, rudd* (Scardinius erythrophthalmus) *top right,*
and tench (Tinca tinca) *below right.*

In Europe and North America, the first freshwater fishes kept in captivity were the local species, and it is with some of these that this series of pictures starts: perch, pike, rudd and tench.

The perch, *Perca fluviatilis*, is common throughout most of Europe, although rare in Scotland and Norway. In North America its place is taken by the closely related, but smaller, Yellow perch, *Perca flavescens*.

For a European fish, the perch may seem to be strikingly coloured when away from its environment. In its own environment, however, the dark vertical bars on the body function as a disruptive colour pattern, thereby concealing the perch's presence from its prey. Perches are commonly shoaling fish, and within the shoals the fishes are approximately the same size. The very large perch (about four to five pounds in weight) are more likely to be solitary. Perches tend to prefer still or gently flowing water; in faster flowing rivers they will congregate in the eddies.

The colours of the perch vary throughout the season and throughout life. The young are usually paler than the adults and the males more intensely coloured than the females. All are darker in winter than in summer, whilst a courting male in spring is probably the most strikingly marked of all. During winter, perch congregate in the deeps, but in spring they start to migrate into the shallows. They normally breed between March and May (this will depend upon the locality and the weather conditions), usually in areas where there are plenty of reeds or sunken branches. They have to be particular in their choice of breeding site because the eggs of the female are encased in a tough membrane which has to be rubbed against an obstruction until the eggs in the long string are released. The egg strings are fertilised by the male. Anchored at one end, and floating in the current, the eggs hatch in a few days.

Our common perch has lent its name to one of the largest orders of living fishes, the Perciformes. They are advanced fishes, characterised, amongst other things, by spines instead of soft rays at the front of the dorsal fin, "spiny" scales and pelvic fins below the pectoral fins. The spiny scales are given the name "ctenoid", meaning comb-like, which refers to the serrated posterior edges. If you run your finger along the side of a perch from tail to head, you will be able to confirm this.

The pike (*Exox lucius*), is a solitary piscivore growing to about forty pounds in weight. Because of these two factors, many apocryphal stories have been woven around this fish. Gesner (1558) reported a story of a man taking his thirsty mule to a pond, where a hungry pike bit the lips of the mule so firmly that the mule drew the pike out of the water. There is another story of a woman in Poland who had her foot bitten by a pike as she was washing out her clothes. Only a few years ago, a report appeared in the *Sunday Express* of a giant pike in Ireland, which, when hooked by a boat-fishing angler, towed the boat along the lough before breaking the line.

The longevity of the pike has been the substance of many, often-repeated stories, of which the most famous is the "Emperor's Pike". Again, the story stems from Gesner, who recorded that a pike captured in a Württemburg lake in 1497 had, around its gill region, a copper ring inscribed to the effect that the pike had been placed in the lake by Frederick II in 1230. This pike was nineteen feet long and weighed 550 pounds. Its skeleton is preserved in the cathedral at Mannheim. However, discrepancies soon appeared in the books written by contemporary authors: there was disagreement about the length, date of capture, and which Frederick was allegedly responsible. The myth was finally shattered in the last century when the skeleton at Mannheim was examined. Although there was only one head, there were enough vertebrae to reconstruct several pikes!

Esox lucius is found throughout Europe, Russia and North America. It rarely achieves a weight of more than forty pounds, but a close relative, *Esox masquinongy* from North America, has been reliably reported to reach one hundred pounds in weight and six feet in length. While there is only one species of pike in Eurasia, there are four in North America, the two larger species mentioned above and two small ones called pickerels, *Esox americanus* and *Esox niger* which rarely exceed ten pounds.

Incidentally, the fish known to Americans and Europeans as the pike-perch is not a hybrid but an elongated relative of the perch. It has the spiny dorsal fin and ctenoid scales of the Percidae, not the smooth scales and soft-rayed dorsal fin of the pike.

The rudd (*Scardinius erythrophthalmus*) is a cyprinid fish, often confused with the very similar-looking roach. They can be distinguished with a little care: the former has eight or nine branched rays in the dorsal fin and an inclined mouth, whilst the latter has nine to eleven branched rays in the dorsal fin and a horizontal mouth. Much to the surprise of many anglers, rudd are a commercial food-fish in eastern Europe where they are caught in nets.

The tench (*Tinca tinca*) is another cyprinid, once valued greatly as a source of food by monks, who used to keep them in the monastic stew ponds. Tench have a firm, white, or slightly brown, flesh, the quality of which varies with the locality. The "muddy" flavour which may be present (although not necessarily in tench from muddy waters) is apparently removed by scalding, which takes the slime from the body.

Country folk used to call the tench the Doctor Fish, for it was said that an injured fish would rub itself against the tench's slime which acted as an ointment. Juliana Berners wrote: "A Tenche is a good fyssh: and heelith all manere of other fysshe that ben hurte yf they maye come to hym." In particular, the pike was supposed never to eat the tench, presumably out of a respect for its healing powers. Unfortunately, modern pike, unaware of this neutrality, will happily eat tench. Observations have certainly been made, by most reliable witnesses, of individuals or small groups of fish rubbing against a tench, but the significance of this behaviour is obscure. One is also informed that a tench, laid upon the liver, is a remedy for jaundice and that tench on the hands and feet will cure fever.

Plate 1

Courtship and parental care. Three-spined sticklebacks (Gasterosteus aculeatus) and their nest.

The Three-spined stickleback, *Gasterosteus aculeatus*, is probably most peoples' introduction to fish, fishing and fish-keeping. It is a common, widespread, hardy little fish, found in nearly every body of water throughout almost the entire northern hemisphere. It is equally at home in fresh water and in the sea, although it will rarely be found more than a few miles offshore. Nevertheless, it is a puzzling fish, full of surprises for both scientist and layman. Its complex courtship has been studied by animal behaviour experts, the changes in its body armour in different waters by physiologists and anatomists, and even now we do not know which group of fishes is most closely related to the stickleback.

In Britain, the Three-spined stickleback was given a bewildering variety of local names; prickleback, spricklebag, branstickle, brandie, sharpie, hurry-bannings, stand, Jack Sharp, tittlebat, fiery loch, enemy chit and so on. Although it is usually just called the stickleback now, the male, when in breeding colours, is still known as redthroat or hackle.

The stickleback is a small fish. Until a few years ago, a specimen four inches long was regarded as a giant, but recently a population has been found living in a lake on an island off the coast of British Columbia, in which individuals reach over nine inches. Throughout the whole of its range, the stickleback shows considerable variation in certain characteristics, so great, in fact, that at one time there were thought to be several species. There are no scales; instead there is a series of bony plates or "scutes" along the side of the body. It is generally, but not universally, true that sticklebacks from brackish or salt waters have more and larger scutes. Some freshwater populations have no scutes at all. In the most exaggerated condition, the scutes have median ridges and a pair of lateral keels are developed in front of the caudal fin. Although called the Three-spined stickleback, specimens are known with two, four or five spines in front of the dorsal fin. The pelvic fin is represented by a strong spine and one weak ray, yet in some Canadian populations the pelvic fin is absent. In the presence of this amount of variation it is not surprising that the fish has always fascinated naturalists.

Even in these days of polluted waters, it is still a common fish, but not as abundant as it was a few hundred years ago. Thomas Pennant, writing in 1776, noted that huge shoals would come up the river Welland every few years. The shoals were so large that they were caught for use as fertiliser and a man employed to catch them at a halfpenny a bushel could earn four shillings a day.

The stickleback may be small, but its voracity and pugnacity cannot be questioned. William Arderon (*see* Introduction) noted that "on the 4th May, one devoured in five hours seventy-four young dace which were a quarter of an inch long and the thickness of a horse-hair. Two days afterwards it swallowed sixty-two and would probably have eaten as many every day as they could have procured for it." In 1879, Francis Day reported that Mr Mable, of the Weston-super-Mare Aquarium, added some roach to a tank containing sticklebacks. The prior inhabitants were dissatisfied, but not frightened by this invasion, for they promptly attacked the roach. The sticklebacks would position themselves in front of the roach, dart forwards and bite pieces out of the roaches' lips. The attacks continued until the roach were killed, and then eaten, by the sticklebacks.

This pugnacity is particularly noticeable when the male stickleback is defending his territory. Outside the breeding season, sticklebacks live in small, moderately harmonious groups. In spring the males leave the groups, develop their red chests and begin to claim territories. Each male defends his territory, especially against other males in breeding dress. When another male approaches, the defender darts forward with his dorsal spines erect and his mouth open. Often, the threat is enough and no fight ensues. If, however, the invader is not frightened, the defender changes his posture and hangs, head downward, with both pelvic spines raised and jerks towards the bottom of the pond. This situation will usually end in a fight, probably near the edge of the territory.

When the territory is established, the male builds a spherical or cylindrical nest from pieces of waterweed glued together by a secretion from his kidneys. Now the courtship starts. The colours of the male become intensified, red on the underside, bluish on the back. The silvery females are swollen with eggs and when one swims near to a male's territory he tries to lure her into the nest. He swims in a zigzag course around the female and if she is ready she follows him. The male turns on his side and pushes his head through the side of the nest. The female wriggles into the nest with her head and tail protruding at either end and the male encourages her to spawn by butting the base of her tail with his snout. After she has spawned she departs. The male then enters the nest and fertilises the eggs. When two or three females have laid their eggs, the male's sexual drive is replaced by a brooding phase. For about a week, until the eggs hatch, the male fans the eggs with his pectoral fins to ensure their aeration. When the young hatch out he plays "sheepdog", retrieving stragglers by catching them in his mouth and spitting them back into the nest. In a fortnight to three weeks, the young fishes will start to form shoals. When they do this, the male loses his breeding dress and leaves them to fend for themselves.

The sticklebacks' reactions have been studied in some detail. For example, a male will attack a crude model with a red underside more violently than an accurate model that is not red. Each reaction of the stickleback is involuntary and the reactions always follow the same sequence, for example the innate releasing mechanism involved in establishing a territory is replaced by a nest-building phase and so on.

There are other species of sticklebacks, but they have been less intensively studied. The Fifteen-spined stickleback, *Spinachia spinachia*, is entirely marine, and is found in shallow seas around the European coast. It grows to about seven inches long and makes fist-sized nests in rock pools. The Ten-spined stickleback, *Pungitius pungitius*, is smaller and less well armoured than the three-spined.

The main function of the spines on the back and pelvic fins is thought to be defensive. Certainly, birds have occasionally been found dead with a stickleback stuck down their throats, but the mechanism is not foolproof because trout, fish-eating birds and otters are all willing and able to consume large numbers of these spiny fish.

Plate 2

The South American aruana (Osteoglossum bicirrhosum) *above, and a* mormyrid (Gnathonemus petersii) *from Africa.*

Mormyrids and osteoglossids are primitive fishes. They are related, both being contained within the superorder Osteoglossomorpha (which also contains the African Butterfly fish shown in the next plate). Fossilised relatives of the osteoglossids are found in deposits of the Eocene period in the London clay and in North America.

Today only six species of osteoglossids are left; three live in South America, while northern Australia, Africa and New Guinea have one species each. The osteoglossids include among their number a candidate for the title of the largest freshwater fish, *Arapaima gigas*, the pirarucu, or arapaima. Fittingly, the arapaima lives in one of the largest river systems, the Amazon. Its maximum size is not known with certainty, but lengths of about fifteen feet have been quoted. There is, however, an unfortunate tendency for unsubstantiated travellers' tales to elongate large animals, so possibly a maximum length of about ten feet may be more accurate. The rate of growth of an arapaima is in accordance with its size. A specimen kept in the London Zoo aquarium grew from six inches to six feet in six years. The other species are not so large; they may reach a length of three feet.

As befits their size, all the species, except for the African *Heterotis*, are predators; indeed, it is said that the favourite bait for the arapaima is its cousin, the aruana. *Heterotis* has a small mouth and feeds largely on insects, crustaceans, molluscs and seeds. The aruana, *Osteoglossum bicirrhosum* is predominantly carnivorous when adult, but the young (shown here) eat mostly insects. The two leaf-like tentacles on the lower jaw are extended stiffly in front of the fish, touching the surface of the water. The vibrations from any insects struggling in the water are received by the tentacles and the fish has its meal. The tentacles are relatively much smaller in the adult, which is capable of eating larger prey.

The archaic nature of osteoglossids is detectable not only in their superficial appearance, although that alone is enough to suggest a comparison with the early fishes. Within their skull bones there is a peg-and-socket device whereby paired bones of the palate (the endopterygoids) articulate with a median bone (the parasphenoid) running along the underside of the braincase and orbital region. The name "osteoglossid" means bony tongue, and in these fishes the principal bite is between the tooth-bearing bony tongue and the teeth on the parasphenoid. In almost all other fishes, the bite is between the upper and lower jaws at the sides, not the centre, of the mouth.

However primitive they may be osteoglossids make good parents. *Heterotis*, for example, constructs a nest in shallow swamp water. The parent clears an area about four feet across and uses the plants thus removed to make a wall around the nest. The eggs are laid on the cleared swamp bottom. *Osteoglossum* and *Scleropages* are mouth brooders. They keep the fertilised eggs and then the young in the mouth until the young can fend for themselves. Both arapaima parents help to scoop out a hole in which the eggs are laid. The eggs hatch in five days and the young are protected by the male for a further forty days. The male has a gland on his neck which exudes a chemical (a pheromone) to which the young are irresistably attracted. Wherever the parent swims, the young follow, clustered around his neck. By the time they are ready to relinquish his neck, they are about four inches long.

Mormyrids are found only in Africa, where more than 150 species are distributed from the Nile to South Africa. The greatest concentration of species, however, is in the Zaire basin. The species illustrated, *Gnathonemus petersii*, indicates the justification for the common name often applied to this group – Elephant trunk fish. In *G. petersii*, the chin is prolonged as a fleshy sensitive "finger", while the mouth remains in the normal position. Species of the genus *Campylomormyrus*, for example, have the snout extended into a long downward-curving tube which may be more than a third of the body length. In this case, the small mouth is at the end of the tube. Not all of the mormyrids have elongated snouts. On the contrary, *Petrocephalus* species have an almost bulldog-like appearance.

In spite of these extreme variations in mouth form, a mormyrid is always recognisable as a mormyrid. They have smoothly-scaled bodies that seem to be mucus-covered, soft-rayed dorsal and anal fins usually set opposite to one another and a delicate, forked tail fin supported on a slender base (the caudal peduncle). The mouth is small, with a few weak teeth.

Mormyrids have poor eyesight and are very frequently found in muddy waters. How then are they able to feed, find mates and defend themselves? Part of the answer rests with their brain, their mucus-like covering and their weak electric organs. Mormyrids have extremely large brains, between about one fiftieth and one eightieth the relative weight of their body weight (this is more than twenty-five times the relative weight of a pike's brain, and approaches the human ratio of about one twenty-fifth to one fortieth). The cerebellum, in particular, is greatly enlarged, and this is thought to be related to the perception of the fishes' electric field. In the caudal peduncle there are paired electrogenic organs which create an electric field around the fish. If the fish approaches an object, or if the fish is approached, the resultant disturbance in the electric field is detected by special organs (mormyromasts) in the skin. The mucus-like layer is a modified epithelium perforated with pores leading to the mormyromasts and is believed to aid the detection of changes in the electric field. The pattern of electric pulses and the rate of discharge varies from species to species. The presence of mormyrids in a river can be detected from their pulses by placing electrodes in the water and connecting them to an amplifier. The electrical discharges are used entirely for location purposes and not for offence or defence.

Mormyrids are capable of other surprises too. The very long-snouted species look extremely cumbersome and it had been thought that they inhabited still or very gently flowing waters, their long snout probing into crevices for food. It is now realised that these species inhabit rapids and cascades. It seems that they are less concerned about their hydrodynamic problems than we are!

Plate 3

Freshwater Flying fishes. The South American Hatchet fish (Gasteropelecus levis) *top right, and the* African Butterfly fish (Pantodon buchholzi).

Within this book there are burrowing fish, walking fish, electric fish, nest-building fish, blind fish and transparent fish. This plate concerns flying fish.

To many people, the Flying fish evokes dreams of warm weather, and of ships sailing across a clear blue ocean. Suddenly, a silvery shape leaves the upslope of a white-topped wave, reaches a height of a few feet, glides and descends. Occasionally, a high-flying Flying fish will have the misfortune to land on the deck of a ship at night. In the morning the curious will find a fish rather like a herring with greatly expanded pectoral fins. These deck deaths were popular with sailors. According to Captain Hall at the beginning of the last century "we used to pick up Flying-fish enough about the decks in the morning to give us a capital breakfast. They are not unlike whitings in taste, though rather firmer, and very dry." The marine Flying fishes have been known for a long time. Oppian, the post-Aristotelian natural historian wrote of them in his 'Halieuticks of the Nature of Fishes', a book more famed for the rhyming couplets of the first English translation (1722) than for the observations. He wrote:

> "With strange surprize we view the dubious sight,
> Of Fish in Shape, and yet of Birds in Flight,
> *Sea-Swallows* lower fly, regard the Main,
> Mount in their Fear, but quickly dive again,
> But cautious *Hawks* tho' wing'd, will nearer keep,
> And hov'ring o'er the wavy Surface sweep,
> They rince their moisten'd Wings, as close they skim,
> Both Elements enjoy, and flying-swim."

An Appendix at the back of the 1722 edition of 'The Halieuticks' identifies both "Sea-Swallows" and "Hawks" as Flying fish.

There is a surprisingly large number of species of oceanic Flying fish but a general description would be of a silvery fish with a cigar or herring-shaped body, six to eight inches long with large, wing-like pectoral fins. Characteristics not present in all species are expanded pelvic fins (giving a bi-plane effect) and an enlarged lower lobe of the caudal fin. The major fallacy is that these Flying fish fly. They do not fly, they glide.

To gain some idea of how this gliding may have evolved, an examination of the oceanic Flying fishes' relatives gives some clues. There are three closely related groups – the Half-beaks (included in the same family as the Flying fish, the Exocoetidae), the garfish (Belonidae – famous for the green bones of *Belone belone*) and the skippers (Scomberesocidae). The last family gains its common name from the habit of skittering across the surface of the water. The front part of the body is in the air, the posterior part is vibrating rapidly, providing propulsion and controlling the angle of the body. Some garfish have been seen to behave in the same way with the body held at an angle of thirty degrees to the horizontal. If such a fish were to have a pair of wings for gliding, would take-off occur? The answer seems to be yes. An Indian Ocean Half-beak has pectoral fins much longer than all the other species and with a sufficient skittering speed takes off and glides for twenty yards or so.

The gliding of the oceanic Flying fish is perhaps more understandable in the light of these observations. The process varies somewhat from species to species, but as a generalisation may be described as follows: the fish increases the swimming speed up to about twenty miles per hour and starts to come out of the water. The angle between the body and the water is increased and the pectoral fins are opened to provide lift. The tail is still in the sea and vibrating about forty to fifty times a second. Momentum is increased and as the fish is facing into the wind (already using the slope of the waves), it leaves the water with a final thrust from the caudal region. Glides can last for a distance of half a mile, and, with a suitable wind, can reach over thirty feet above the sea. Why do they do it? It seems that the main reason is fright, especially in the presence of a predator.

Unlike the oceanic Flying fishes, the two freshwater Flying fishes shown here actually fly, their pectoral fins vibrating rapidly and thus prolonging the flight.

The deep chest of the South American Hatchet fishes houses the powerful muscles necessary to pull down the pectoral fins, thereby giving uplift to the fish. Hatchet fish do not seem to need a long "run" to take off. When some examples were placed in a jar while being studied by a scientist, one of them suddenly hit the investigator on the wrist over two feet above the water surface. It was thought likely that the buzzing noise that accompanied this feat was the rapid vibration of the pectoral fins aiding the momentum of the initial leap. Observations in the wild report Hatchet fish flying through the air for up to fifteen feet and three feet above the water. This usually happens to escape from predators.

The African *Pantodon* behaves in much the same way. Although the chest is not so deep, the fish has a thicker body and the large pectoral fins are well equipped with muscles to provide a powerful down stroke. The morbid anatomy of the freshwater Flying fishes is known, but the actual mechanics and nature of the flight are still under investigation. High-speed film techniques are being used to find out what really happens.

In the meantime, aquarists owning any gasteropelicines or *Pantodon* should remember to keep a cover on the tank.

Plate 4

Barbs from Asia. At the top the Black ruby (Barbus nigrofasciatus) *from Sri Lanka, on the right the Checker barb* (Barbus oligolepis) *from Sumatra, and to the left the Spanner barb* (Barbus lateristrigata) *from Malaysia.*

This, and the following eight plates, are concerned with the dominant group of freshwater fishes, the Cypriniformes or Ostariophysi. Cypriniform fishes are found in almost all freshwater bodies of all the continents except Antarctica (which has no permanent fresh water, at least in liquid form). The exact number of species is not known, but is well over two thousand – at least ten per cent of all living fish species, and the majority of the freshwater species. The Cypriniformes comprise three major and a few minor groups. The major groups are the Siluroidei or catfishes which typically have teeth and no scales; the Characoidei (characins) which have teeth and scales and the Cyprinoidei which have scales but no teeth. So successful are these groups that it has been estimated that there are over one thousand species of catfish in South America and over six hundred species have been ascribed to one cyprinid genus, *Barbus*.

With this pedigree, it is pertinent to ask what anatomical or physiological peculiarities have enabled the Cypriniformes to be so successful. One undoubted factor is the Weberian Mechanism, named after its discoverer Professor Weber, which is possessed by all cypriniform fishes. This mechanism consists of a series of four small bones, on each side of the first four vertebrae, which link the swimbladder to the inner ear. These ossicles pivot and mechanically transfer any compression waves picked up by the swimbladder to the inner ear. It is a system analogous to the malleus, incus and tripus in our own ears which transfer the sound (= compression) waves picked up by the tympanum to our inner ear. This enables the Cypriniformes to "hear", an asset of great value in muddy waters or at breeding time.

Another factor is the presence of an alarm substance. Within the skin of cypriniform fishes are secretory cells which, when the skin is broken, release the alarm substance into the water. This diffuses through the water, or is carried by the current, and on being smelled by individuals of the same species warns them to flee or hide. With these advantages and a built-in adaptability, it is not surprising to find cypriniform fishes in almost every water condition, from waterfalls to small ponds, from cold to hot, from stagnant to highly oxygenated. Even without teeth in the mouth, the cyprinoids have evolved into many species which range in length from one inch to over five feet.

In this painting we are concerned with three small, colourful Asian members of the cyprinid genus *Barbus*. These three are, top to bottom, the Black ruby, *Barbus nigrofasciatus*, from Sri Lanka; the Checker barb, *Barbus oligolepis* from Sumatra and the Spanner barb, *Barbus lateristrigata* from Malaysia. It has been the habit of several authors to place the small colourful barbs in the genus *Puntius* but as *Puntius* is so inadequately defined this practice cannot be justified. *Barbus* species occur throughout Europe, Asia and Africa. In Britain and western Europe they are represented by the barbel, *Barbus barbus*, a cylindrical fish of fast flowing waters. The name barbel means "little beard" and refers to the four fleshy barbels around the mouth. It is a species loved by anglers for its

sporting qualities, and, although it rarely weighs more than twelve pounds, its fighting spirit makes up for its lack of size. Of the barbel, Juliana Berners wrote: "The Barbyll is a swete fysshe, but it is a quasy meete and peryllous for mannys body. And yf he be eten rawe: he maye be cause of mannys dethe: Whyche hath oft be seen." However, "mannys dethe" is supposed to have benefited the barbel of the Danube. After the carnage of the Turko-Austrian war, so many bodies were thrown in the river and eaten by the barbel that the fishes grew to a large size and were so abundant as to cause many comments.

While there are a handful of *Barbus* species in Europe, over four hundred have been described from Africa. The great majority are small fishes, not usually brightly coloured, living in small streams; on the other hand a specimen of *Barbus kimberleyensis* weighing thirty-seven pounds was caught in a dam in South Africa and *Barbus tropidolepis* often grows to three feet long in Lake Tanganyika. All the *Barbus* species known are egg layers, but Weber (of mechanism fame) described a species from southern Africa as *Barbus viviparus* because he thought he had found developing young in the ovary of a dissected specimen. Quite rightly, Weber regarded this as a remarkable feature for a *Barbus*. However in 1941 the South African ichthyologist, Dr Keppel Barnard showed that Weber was mistaken. The embryos proved, on close examination, to be those of a cichlid. Despite the erroneous reason for the specific name *viviparus*, the name stands, and there are no exceptions, so far as is known, to the egg-laying habits of the cyprinids.

Although some *Barbus* in Africa are large, they are dwarfed by the mahseer of India. It may be more correct to write mahseers for there is dispute as to whether there is one species (usually known as *Barbus tor*) or several. Many people consider the mahseer to be the greatest freshwater sporting fish. It lives in the rivers of the Himalayan foothills where the climate is as excellent as the fishing. As with most sporting quarries, it is difficult to get any precise information. Without a specimen, the size will have increased by the time the report reaches disinterested ears. Equally, it would prove difficult to carry a large specimen (which is usually an unforseeable catch) to a suitable repository where it could be measured. However it is probable that the mahseers reach six feet in length and some authors have quoted nine feet. The scales, which were common souvenirs, are at least the size of a man's palm.

Keeping a mahseer in captivity would seem to be beyond the capability (or desire) of the average aquarist. Nevertheless *Barbus hexagonolepis*, a close relative of *Barbus tor*, is often on sale and in the wild it grows to two and a half feet long.

The genus *Barbus* has achieved many things. In size they range from the mahseer to the small pretty Butterfly barbs (*Barbus hulstaerti* and *Barbus candens*) of Africa, which are only one inch long. Most barbs live in rivers or lakes, yet in Africa there is a blind, white cave-dwelling species. In contrast to those depicted in the plate opposite, the majority are of a uniform, inconspicuous hue.

Plate 5

Cyprinids of South-East Asia. Rasbora maculata *top left,* Rasbora trilineata *bottom right,* Rasbora hengeli *top right, and* Rasbora heteromorpha *bottom left.*

The genus *Rasbora* contains a large number of small, pretty species who have their greatest flowering in Malaysia. They are found, however, from Ceylon through South India to the Indo-Australian Archipelago and from Canton to the Celebes and Philippine Islands. Some books give East Africa as a locality for a *Rasbora* species, but this is probably based on an old error. *Rasbora zanzibarensis* was described by Gunther and Playfair in 1866 as coming from Zanzibar. The description was based on two specimens and no more have ever been found. It is now thought that, prior to Gunther and Playfair's observations, there was a confusion and a jar containing specimens of *Rasbora daniconius* from India became included in the African collection. The knowledge of the distribution of freshwater fish was insufficient in the 1860s to enable the inconsistency to be detected at once. It is astonishing to realise that a hundred-and-ten-year-old mistake is perpetuated.

The characters uniting the various species into the genus *Rasbora* are the infrequent presence of barbels and the existence of a knob at the front of the lower jaw that fits into a corresponding recess at the front of the upper jaw. Some species have an incomplete lateral line, while in others the lateral line is complete. In the wild, many *Rasbora* species live in huge shoals near the surface of both rivers and lakes. Although the individuals are small, the concentration of fish in the shoals is such that they form an important item of diet for people in several parts of South-East Asia.

The other great value of *Rasbora* species is as aquarium fishes. Although there are very many species of small, pretty freshwater fishes in the world, only a small proportion are commonly kept in aquaria. Several factors are involved in determining which species are going to finish up in the local pet shop. Firstly, the fish must live in water within a reasonable distance of the shipping centre. The carriage of fishes from their native habitat to the shipping centre is an expensive business and an arduous process for the local collectors. Secondly, either the species must be abundant, or, if not abundant, must be spectacular enough to command a high price. Thirdly, the fishes must travel well so that a high proportion arrive alive. The motive behind the third factor is not merely altruism on behalf of the exporter, but hard commercial sense because the importer is not going to pay the exporter for dead fish. Fourthly, the public must be sufficiently attracted to the fish and aware that it is not too difficult to keep before they buy it.

Rasboras score heavily on all these counts. They are numerous, easily collected, travel well, are hardy and easy to keep. As a bonus they are pretty (although the majority are not as gaudy as the South American tetras) and several species are easily bred by amateurs.

One such is the Harlequin fish, *Rasbora heteromorpha*, which is found in many localities in the Malay Peninsula, Thailand and parts of Sumatra. This fish was first imported into Europe in the early years of this century, surely a tribute to its hardiness, for the journey would take about a week, which is in sharp contrast to the speed of jet transportation now. In all, eleven species of *Rasbora* are known to have been brought to Europe before the First World War. Not all of these proved successful or popular, but all withstood the journey. To put this into perspective, only five species of the far more speciose tetras (genera *Hyphessobrycon* and *Hemigrammus*) are known to have been imported into Europe in the same period. Extremely similar to the Harlequin fish, but not discovered until 1955 is *Rasbora hengeli* from Sumatra (*Rasbora heteromorpha* was discovered in 1904). The two can be distinguished by the more slender body and more slender, slightly fainter triangle of *Rasbora hengeli*.

The scissor-tail, *Rasbora trilineata*, is one of the larger species. In its native environment (Borneo, Sumatra, Malaysia, Thailand and the Sunda Islands) it grows to over six inches long. The common name derives from the scissor-like closing of the lobes of the caudal fin as the fish starts forward or adjusts its position in the water. *Rasbora maculata* is one of the more brightly coloured species. Although it rarely grows to more than an inch in length it was arguably the first *Rasbora* species brought to Europe.

One of the side effects of the demand for *Rasbora* species within the aquarium trade, is that the collectors in the field will sample new areas from time to time. This, in recent years, has yielded several new species, some of which have appeared in aquarium shops before they had even been recognised as new species! During 1976, for example, two new species of *Rasbora* were described. Both were recognised in the holding tanks of fish exporters, and both species are more spectacular than any previously known.

The first of these, *Rasbora brittani*, is known as the head-and-tail-light *Rasbora*. This species was found in a collection of fish from the Johore river, Malaysia. The body is brassy above and silvery below. Just in front of the tail fin there is a small intense black spot, above and below which are bright orange spots. The headlight part refers to the orange colour at the top of the iris. The longest rays of the dorsal fin are also orange. Although very pretty, it is an insignificant species beside the gaudiness of the Neon rasbora. This species, which comes from Sumatra, is called *Rasbora axelrodi*. The back is grey with gold highlights. Running from the head to the tail is a brilliant emerald green stripe about as wide as the eye (very like the blue/green stripe in the Neon tetra). The chest and belly are bright red. The anal fin has a scarlet membrane with dark brown rays.

It may seem surprising that a fish as spectacular as this can have remained unnoticed for so long. Yet, as the fish is only three-quarters of an inch long and comes from a poorly-known region, it is perhaps understandable. The interesting point is that these two species were discovered by the efforts of commercial fish exporters, collecting specifically for the aquarium trade.

Plate 6

Cyprinid fishes, called "sharks" from South-East Asia. The Red-tailed black shark (Labeo bicolor) *and the Silver shark* (Balantiocheilos melanopterus).

There are real sharks and there are also bony fishes called sharks. Real sharks are marine fishes with cartilaginous skeletons and a leathery skin in which tiny enamelled denticles are embedded. Most real sharks are large (although there are a few deep-sea exceptions which only grow to a foot long) and all sharks have a series of five (rarely six or seven) gill slits behind the head. Real sharks are closely related to the skates and rays which also have cartilaginous skeletons and the other features characteristic of this major group of fish.

The "sharks" which are the subject of this painting are not real sharks but ostariophysans, relatives of the barbs and danios. Quite why their common names should include the word shark is difficult to discover. They do not look like sharks, they do not behave like sharks and they are about as closely related to real sharks as an ostrich is to a duck-billed platypus. Nevertheless, most English-speaking aquarists refer quite happily to the Red-tailed black shark and the Silver shark. These two species are not the only "sharks". A catfish from South-East Asia, *Pangasius* species, is commonly called the Blue "shark" (definitely not to be confused with the real Blue shark, *Prionace glauca*). Apollo "shark" is an epithet applied to the innocuous cyprinid, genus *Luciosomus*, from South-East Asia. The black "shark" (*Morulius chrysophekadion*), also from South-East Asia, is a close relative of the Red-tailed black shark. The only apparently common factor among the ostariophysans with the epithet "shark" is that they come from South-East Asia.

The Red-tailed black shark (*Labeo bicolor*) is naturally black with a red tail. In many cases, bizarrely-coloured fishes have been specially bred in captivity in order to produce the apparently unnatural colours. But *Labeo bicolor* needed no such assistance from breeders. In Thailand it is known by two names, pla song kruang (Full-dress fish) or, more mundanely, pla hang deng (Red-tail fish).

The genus *Labeo* occurs in Africa and throughout South-East Asia. All are vegetarians or bottom feeders, and they have soft, expanded lips which may bear harder papillae, or ridges, to assist in scraping food. Generally, the African species are rather drably coloured, shades of olive green or brown are common. But the Asiatic *Labeo* are more brightly coloured with several species possessing a red tail. Admittedly, *Labeo bicolor* is the most striking example, but the tendency is seen in other species. *Labeo erythura*, for example, (arguably this species could be the same as *Labeo frenatus*) has a brown body with an orange-red tail. *Labeo munensis* has large, conspicuous black dorsal, anal and pelvic fins, all with a sharply-defined white edge, but the tail is white.

So far, no satisfactory explanations have been proposed to explain the advantage to the fish of the sharply contrasting tail and body colours. In many fish, striking colours serve to warn potential predators that their owner is poisonous or has a nasty taste. This is not true for the small *Labeo* which are contentedly eaten by larger fish. Perhaps a theoretical consideration of this problem will be less beneficial than admiring the beauty of *Labeo bicolor* as it swims around the tank. Who knows, maybe a chance observation by an aquarist will provide the answer.

Whereas the Red-tailed black shark only grows to some five inches long, the Silver shark *Balantiocheilos melanopterus* grows to over a foot long. *Balantiocheilos* occurs in Sumatra and Borneo as well as on the mainland, but the largest examples come from the islands. The Silver shark, with the grace and elegance of its body and the black edges to its fins, is one of the most subtly beautiful cyprinids. The black-edged fins are commemorated in a local name from central Thailand – pla hang mai or Burnt-tail fish. Like many cyprinids, this species is an omnivore; it seems to prefer small live food but will thrive on detritus and algae as well. Although they are an expensive addition to an aquarium in the western world, in some parts of Thailand they are extremely common and large shoals are netted by fishermen. The fish is not regarded highly as food and is frequently converted into a stew.

Only small specimens are kept by most aquarists. From a naturalist's viewpoint, this is a pity, because one attribute of the fish will not be seen; and that is its prodigious powers of leaping. The leaps occur all the time, not merely during the spawning season, and there is a record of a *Balantiocheilos* performing a vertical leap of over six and a half feet. This, for a foot-long fish, is the equivalent of a high-jumper clearing a bar set at forty feet.

Whilst we are considering species of fish from South-East Asia, for interest, we cannot exclude *Leptobarbus hoevenii*, another cyprinid that has been imported into Britain and America during recent years. The most fascinating aspect of the behaviour of this species is again not likely to be seen by aquarists. The Thai name is Pla ba which means Mad fish. *Leptobarbus* is particularly fond of the fruit of the chaulmoogra tree (*Hydnocarpus*) and when the rains wash the fruit into the river, the fishes gorge themselves. Unfortunately, the flesh of the fruit has an intoxicating effect on the fish which, after a while, swim about wildly and erratically, displaying symptoms analogous to those of an enthusiastic toper. Although *Leptobarbus* has a poor reputation as a food fish under normal circumstances, it is regarded as poisonous when under the influence of the chaulmoogra fruit.

Plate 7

The Veiltail goldfish (Carassius auratus).

The goldfish (*Carassius auratus*) is almost certainly the species which has been kept and cultivated for its beauty longer than any other species. Although the goldfish is now found in almost as many countries as the guppy, its native home is South-East China and possibly some of the off-shore islands. There, in its wild form, it is an inconspicuous brownish fish, closely resembling its relative, the Crucian carp, *Carassius carassius*.

Two factors have contributed to its long association with man: the first is that the goldfish is a hardy beast and can tolerate a wide range of water temperatures and conditions. This meant that it took kindly to captivity. The second factor is that some individual wild goldfish, because of a not-infrequent genetic mutation, occur in red or gold livery. The olive-brown colour of the wild goldfish is produced by light passing through a mixture of black pigment cells (melanophores) and orange-coloured pigment cells (chromatophores). A loss of the black cells will result in an orange-red coloured fish called a xanthochromic individual. A loss of all the pigment cells will produce a white, albinistic, individual. The ancestral stock of domesticated goldfish were xanthochromic individuals occurring in the wild.

This tendency is not limited to goldfish: Golden tench and Golden orfe, which are also frequently kept in aquaria, are produced by a similar loss of black pigment. A population of Golden roach has been known in one area of the river Tyne since the 1930s, so the characteristic may be hereditary. Rarely, Golden pike are found and in Africa the occasional Golden Nile perch is caught. In recent years a xanthochromic variety of a loach has been on sale in shops, and the common Golden barb is well known to aquarists. This list is by no means complete, but merely illustrates the fact that the same phenomenon occurs in unrelated groups of fishes.

The Chinese recorded the presence of wild fishes with red scales over 1600 years ago. During the Chin Dynasty, Huan Ch'ung (328–384) noticed them in a lake during a visit to Mount Lu. The first unchallenged evidence of the maintenance (and therefore breeding) of goldfish in captivity occurs in the Sung Dynasty. About 1030 the 'Poem on the Pagoda of the Six Harmonies' was written. There, the poet remembers waiting by the Pine Bridge for the goldfish (Golden Chi). (The Pagoda of the Six Harmonies, incidentally, was built near Hangchow in 971. It has been restored several times and probably still remains in much the same condition as it was when the poet waited for the arrival of the Golden Chi.)

The isolation of China from the West delayed European awareness of the goldfish. Possibly the first reference in a European work was by M. Martini ('Description Géographique de l'Empire de la Chine') who lived in Hangchow for a few years before 1650. He recorded "little gilded fish, called by the Chinese Chin-yu because the skin glistens being seemingly interwoven with threads of gold – and the whole back apparently sprinkled with gold dust . . . The Chinese value them highly".

By 1742, goldfish were sufficiently well known for Thomas Gray to write his 'Ode on the Death of his Favourite Cat – Drowned in a tub of Goldfishes'. There can be no doubt that the "Pensive Selina – demurest of the Tabby kind" is not the only cat to have met its fate in such fashion, because:

> "What female heart can gold despise ?
> What Cat's averse to fish ?"

Gronovius in his 'Zoophylacium' of 1763, lists eleven different varieties of goldfish, including forms that approximate to the comet, fantails and veiltails of today, and what could also be interpreted as a lionhead.

The early history of the fancy breeds is very confused and scattered; but in two Chinese works of *c*. 1590 (one of which is the famous 'Book of Vermilion Fish') there are fairly precise descriptions of round-bodied fish, three- and four-tailed fish (each lobe counts as a tail in this context) and possibly a telescope-eyed fish. There is also an important comment which notes that these fish did not exist during the Hung-ch'ih reign (1485–1505); so we may conclude that the sixteenth century was the period of their development. From history back to natural history. The soft rays of the fish's fin are formed of two lateral halves, the base of each half articulating each side of a basal supporting bone. The divided tail fin and anal fin seem to develop as a corollary of the shortening of the body. The simplest phase is where the lower lobe of the tail fin is divided, each part having one of the lateral halves of each fin ray. This can occur as a natural mutation that also produces a slightly stockier body. The interbreeding of such forms produces an even more rounded body with both lobes of the tail doubled (although they may remain joined at the top). A continuation of the same process will split the anal fin into its two lateral halves. In the extreme condition the basal supporting bones of the fins concerned are also separated into their two component lateral halves. It seems rather as if, when the body shortens, the backbone and its processes have split the bones supporting the tail and the anal fin in two, almost as if the tail and anal fin had been jammed into the body and cleaved by the processes of the backbone. The absence of the dorsal fin in some varieties appears to be genetically quite unrelated to the chain of events described above.

The Koi carp, a brightly-coloured fish which has recently become popular, must be mentioned briefly. It is not a goldfish, but a common carp (*Cyprinus carpio*) which has undergone, under the influence of oriental breeders, a process of cultivation similar to that of the goldfish. The origins of Koi breeding are less ancient than that of the goldfish, so possibly in a few hundred years' time a wide range of body form of Koi will be produced to rival that of the goldfish. We must wait and see.

Plate 8

The Clown loach (Botia macracantha) *above,*
and the Coolie loach (Acanthophthalmus semicinctus).

The loaches (family Cobitidae) are small freshwater fishes related to the carps. They have barbels, their scales are greatly reduced or absent and the anterior part of the swimbladder is encased in a bony capsule. They are bottom-dwelling fishes found in a wide variety of habitats throughout Europe and Asia but, unlike their relatives the carps, they have made little impact on Africa. The only African species, *Noemachilus abyssinicus* from Lake Tsana in Ethiopia is known from a unique specimen collected in 1908.

There are two species in Britain, the Stone loach *Noemachilus barbatulus* and the rare and localised Spined loach *Cobitis taenia*. These two are joined by a third species in western Europe, the weatherfish, *Misgurnus fossilis*. The number of species increases eastwards, and the greatest number occurs in the warm waters of India and South-East Asia.

The European loaches are not as colourful as the two species shown in the painting opposite (*Acanthophthalmus semicinctus* – the Kuhli or Coolie loach from the East Indies and the Clown loach *Botia macracantha* from Borneo and Sumatra), but they illustrate many of the characteristics typical of the family Cobitidae.

The weatherfish is a living barometer – hence its common name. In Austria they were formerly kept in jars of water to give advance warning of the approach of thunderstorms. As the atmospheric pressure rises, the fish becomes agitated and frequently rushes to the surface of the water. This unusual activity is caused by the sensitivity of the bone-encased part of the swimbladder. There is a canal which runs from near the pectoral fin, through the bony capsule to the swimbladder, and any atmospheric pressure changes are transmitted via this canal to the swimbladder, thence to the inner ear via the Weberian mechanism.

The Stone loach has not achieved the fame of the weatherfish as a barometer, but it reacts in the same way towards pressure changes. Jonathan Couch, the nineteenth-century naturalist, reported that "when kept in an aquarium it would throw itself over the walls of its prison at the approach of, or during, any extreme changes in wind or weather." Francis Day follows this statement by observing that "they commit suicide by springing out of streams or rivers".

Although a small fish, rarely exceeding five inches in length, the Stone loach has long been considered a delicacy. Dr Rutty in his essay 'Towards a Natural History of Dublin' remarks that "it is customary with many to take it alive with a glass of generous wine." Izaak Walton pronounced it a delicate dish at table and readers of 'Lorna Doone' may recall John Ridd fishing for them with a three-pronged spear.

The name Spined loach refers to the presence of an erectile spine below and in front of the eye. This spine can be locked in an erect position and is supposed to act as a deterrent to predators. As a defence it is a double-edged sword because, although predatory fishes and birds have been killed by swallowing a loach with spines erect, the loach has also died.

Most fishes use their gills for respiration, but a few can breathe atmospheric air by means of an accessory breathing organ in the gill chamber (e.g. the gourami) or by means of a modified swimbladder (e.g. the osteoglossids). Some species of loach, however, use their intestine for breathing air. The Chinese weatherfish (*Misgurnus anguillicaudatus*), for example, breathes air during the summer when the water becomes deoxygenated, but in winter it uses its gills because the water is richer in oxygen. The air-breathing of an Indian loach (*Lepidocephalus guntea*) has been carefully observed. The fish comes to the surface and gulps in air, then turns a somersault and forces the spent air out through its anus in a series of small bubbles, sometimes with enough force to produce a tinkling sound. This peculiar behaviour is made possible because a part of the intestine has a highly vascularised wall through which oxygen can be absorbed into the blood and the carbon dioxide excreted into the lumen of the intestine. In the Chinese weatherfish, the "lung" part of the intestine resumes its normal digestive function when the fish reverts to gill-breathing in winter.

The European loaches exemplify the most common body shape in this family, eel-like with short dorsal and anal fins close to the tail. The genus *Botia* a member of which, the Clown loach, is shown here, comprises species which are much deeper-bodied. They tend to be shoaling species and far less secretive in their habits. The Clown loach, in the wild, grows to more than a foot long. The Kuhli loach is more serpentiform than most loaches and grows only to about four inches. There are several very similar species within the genus *Acanthophthalmus* which are distinguished by the extent and nature of the dark saddle-shaped markings on the back.

A recently discovered loach, *Noemachilus smithi*, is probably the most interesting of all, for it is the only known blind cave loach. (The only one of the thirty-five or so species of blind cave fish commonly kept by aquarists is shown in plate 12 – the Blind cave characin.) The majority of cave fishes are ostariophysans, but until *N. smithi* was found, no loaches had been known to have adapted to a cavernicolous existence. In view of the cryptic habits of loaches, this had been thought a surprising omission. The discovery was by chance. What were thought to be two specimens of a known blind cave fish *Iranocypris typhlops* were brought alive to London from a cave in the Zagros mountains. Later, when they were examined, it was realised that one of the fishes was a loach, pure white and eyeless. Serendipity should never be undervalued in scientific research.

Plate 9

The Electric catfish (Malapterurus electricus) above, and a South American Sucking catfish of the genus Hypostomus.

Catfishes are found in all continents except Antarctica. They are the only ostariophysans found permanently in the sea, where they are represented by two families, the Plotosidae and the Ariidae. Catfishes have barbels but do not have scales. In some the skin is naked, in others the whole body is covered with bony plates. Catfishes are found in all water conditions between subterranean streams and high-altitude torrents. In size they vary between the tiny *Corydoras pygmaeus* of South America and the giant *Pangasianodon gigas* of Thailand which grows to ten feet long but is toothless and feeds on algae. Catfishes include carnivores, vegetarians, omnivores and parasites. In breeding habits there are nest builders, bubble nest builders, oral incubators and those who do nothing but lay and fertilise the eggs. Generally speaking, catfishes are a successful and adaptable group of fishes.

In this painting we are concerned with two species; the Electric catfish of Africa, *Malapterurus electricus* and one of the South American sucking catfish of the genus *Hypostomus*. There are two major families of armoured catfishes in South America, the mailed catfishes Callichthyidae and the sucking catfishes, the Loricariidae. Apart from the mouth shape, the two families can be distinguished because the Callichthyidae (*see* next plate) have two rows of scutes along the side of the body, whereas the Loricariidae have three or four rows.

The loricariids live on algae which they graze from rocks and plant leaves. Usually, but by no means universally, they live in fast-flowing streams. The lips are greatly expanded forming a sucker around the mouth. This sucker (aided by the depressed shape of the body) holds the fish stationary on the bottom of the river while the feeble teeth scrape away the algae. The sucking catfishes do not appear to have any means of offence, but the body is encased in jointed bony plates which act as a defence against predators. The plates are reduced or absent on the underside of the fish. There is supporting evidence about the value of the armour against predators. One of the loricariids, the Capitaine, *Astroblepus*, is adapted to live in torrents and can even make its way up the vertical walls of potholes or waterfalls. In this environment it is safe from predation and, presumably as a consequence, the skin is naked. Even the adipose fin, which is supported by a spine in the armoured loricariids, has lost the spine.

The alimentary canal is unusual in sucking catfish. To start with, they can use the "stomach" as an accessory lung, much as the intestine is used in the weatherfish (plate 9). Secondly, the posterior part of the intestine has a "spiral valve". This is a device, rarely found in bony fishes but common in sharks, in which the lumen of the intestine is filled with a longitudinal spiral of tissue thereby greatly increasing the area available for absorption of food without increasing the length of the intestine.

Most species in the family are shaped like the *Hypostomus* illustrated, but an outstanding variant is found in the genus *Farlowella*, where the fish presents a successful imitation of twig. The genus *Xenocara* is shaped like *Hypostomus* apart from the snout. The male *Xenocara* sport a series of long branching "bristles" on the snout; the females are also whiskered but less emphatically.

The armoured body of the loricariids is, no doubt, of great use against fishy predators but appears also to be of great use to human predators. The Amerindians are fond of loricariid flesh and shoot them with arrows. It appears that the armour plates render the use of cooking pots unnecessary as the fish can be roasted whole and then cracked open.

The Electric catfish is widely distributed throughout Africa from the Nile to the Zambezi. The skin is naked and the general appearance of the fish can forcibly remind one of a soft, bewhiskered sausage. Apart from its shocking qualities, *Malapterurus* can also emit a hissing sound. As befits a fish capable of stunning its prey, it is not an active swimmer, normally hiding away in dark crevices. Despite this, large Electric catfish have been caught with fast-swimming species in their stomachs.

All the other Electric fishes have discrete, well-defined electrogenic organs which, in dissection, resemble translucent, slightly gelatinous muscles. In this case the electrogenic organs form a loose envelope just below the skin but, like all other electrogenic organs, are derived from muscles. A large *Malapterurus* can discharge up to four hundred volts, and the discharge is used for defence as well as for capturing its food.

The development of the electric organs is interesting; basically they develop from muscles. Whenever a nerve instructs a muscle to contract, a small change of electric potential is involved. The electrical stimulus from the nerve is translated into contraction of the muscle fibres at the end plates. The electric organs have developed from the end-plate muscle fibre system at the expense of the contractibility of fibres. With the loss of contraction the end-plate muscle fibre system becomes analogous to a simple battery cell. These "cells" work together, still under the control of the vagus nerve, and can release all the electric potential in one discharge if so desired. There is a permanent electric field around *Malapterurus* which may work in a manner analogous to that of the mormyrids (plate 3).

The peculiar attributes of the Electric catfish have been known for many centuries. Over four thousand years ago, the Egyptians painted pictures of Electric catfish on the walls of their tombs. One surprising observation is that although it is known that the Electric catfish is sensitive to magnetic fields it has been known to become highly agitated several hours before earthquakes. These observations are as yet inexplicable, but the Arabic word for the fish is Raad, which means thunder.

Plate 10

A naked catfish, the Upside-down catfish (Synodontis nigriventris) *top right, and three species of Mailed catfish. In the foreground* Corydoras aeneus, *in the centre* Corydoras julii, *and behind* Corydoras melanistius.

Although of similar body shape, the catfishes represented in this painting come from two continents and show different extremes of body armour and posture.

The four examples of the three species in the foreground are members of the family Callichthyidae from South America. Callichthyids are heavily armoured fishes. The body is enclosed in two rows of overlapping scutes on each side, which do not impair the flexibility of the body. The head is protected by bones immediately below the skin. The fleshy adipose fin is provided with a spine which, in some species, is movable. The leading edge of the pectoral fin is formed from a serrated spine which can be locked erect if the fish is in danger or is actually being swallowed by a predator. In the latter case the death of the predator may result, so also may the death of the catfish.

Obviously, the common name catfish refers to the whiskers or barbels around the mouth. These play an important part in the perception of the environment by the catfish. Most catfishes have small eyes and poor sight, the great majority are nocturnal, or at best crepuscular (although the *Corydoras* species shown here are exceptions), therefore other senses are used to detect food. The barbels of catfish surround the mouth, and on their surface they have a heavy concentration of taste buds. The whiskers act like lashing feelers as the fish moves along and the suitability of food can be determined before the food is in the mouth. The barbels can be moved independently, although the degree of control varies; the maxillary barbels for example are much more mobile than the others. There are commonly six barbels: two nasal, two maxillary (at the edge of the upper jaw i.e. the corner of the mouth) and two mandibular, on the lower surface of the lower jaw. In some species, such as the African *Synodontis alberti*, the maxillary barbels are as long as or longer than the body.

The Mailed catfishes, Callichthyidae are found over much of tropical South America and Trinidad. The *Corydoras* species, so popular with aquarists, are diurnal and can be seen bumbling around on the bottom of slow-flowing forest streams during the day. *Corydoras* species are not confined to a bottom-dwelling existence however. The dwarf species *Corydoras hastatus* and *C. pygmaeus* spend much of their time in mid-water and it has been noticed, in the Orinoco river, that many *Corydoras* species are found in association with floating mats of vegetation brought down by floods.

Callichthys callichthys, the Mailed catfish from which the family takes its name, has a breeding behaviour which is unusual for a catfish, and is not found in the closely-related genus *Corydoras* – it builds bubble nests. The nest is usually built under broad floating leaves and the hundred-plus eggs placed therein are guarded by the male. During the father's sentry-go, he is reputed to make stertorous grunting noises. The oddities of *Callichthys* and its relative *Hoplosternum* do not end there; both these forms have part of the intestine modified into a "lung". *Callichthys* and *Hoplosternum* are often found in swamps, and as swampy water is frequently low in oxygen, the accessory respiratory apparatus is an adaptation to their survival, for it enables them to utilise atmospheric air.

Further, the bony armour (which reduces the rate of desiccation of the body) and the strong spine in the pectoral fin enable *Callichthys* to move overland for short distances when their pool dries out.

Corydoras species have never been reported to leave the water in order to find a larger pool, but their breeding has been observed, and this has several interesting features. One or both potential parents assiduously clean various possible spawning sites. Upon the selection of a very well cleared site, the female sucks at the genital opening of the male. This sucking action draws the male's sperm into her mouth and thence the sperm are carried in the current of water leaving her gills to the eggs, which are now being extruded. During this time the gill covers are opened and closed more vigorously than usual to ensure that the sperm reach the eggs. The eggs are then cradled in a pouch made from the female's ventral fins, after which they are attached to the prepared smooth surface. Cosy as this behaviour pattern may be, it is far removed from the habits of the South American catfish, *Platyastacus*. During the breeding season of this Brazilian fish, the skin on the ventral surface of the female becomes swollen and spongy. When she has laid the eggs and they are fertilised, she sits on them, much in the manner of a hen. However, the eggs sink into her spongy skin and they stay, each egg in its own pit, until the young hatch.

One mundanely normal habit shown by the great majority of catfish is the unadventurous one of swimming the right way up. Not so some species of the African genus *Synodontis*. *Synodontis* species are naked catfish. There are no scutes on the body which, instead, is covered in a thick leathery skin. However the head and gill region are covered with dense bone and the dorsal and pectoral fins bear a strong sharp, serrated spine. There are over eighty species of *Synodontis*, distributed through most of Africa below the Sahara, but only a handful of species swim upside-down.

Fishes are countershaded, i.e. the underside (which would be in shadow with the light coming through the surface of the water) is paler than the back. This arrangement of light and shade helps the fish to be relatively inconspicuous. The upside-down *Synodontis* have either reverse counter-shading (belly darker than back) or are neutrally shaded. The upside-down species spend far more time swimming in mid-water than do many of the normally orientated bottom-living species. The mechanics of upside-down swimming have recently received some attention and certain interesting observations have been made. The adipose fin (behind the dorsal fin) is, as its name indicates, entirely fatty, and fat is lighter than water. Yet the upside-down species have a much larger adipose fin than the right-way-up species. This would seem to have the effect of turning the fish the right-way up, but instead it seems to act as a keel to stabilise the fish. It has been suggested that the inverted swimming might be a response to exploit a food source not normally available to *Synodontis* – food upon the surface of the water. There is no clear-cut division between upside-down and right-way up, some species even spend a proportion of their lives on their side, whilst others can facultatively occupy either posture. Incidentally, a species from another family, *Leiocassis siamensis*, from Thailand, will also swim either way up.

Plate 11

Blind cave characins (Anoptichthys jordani) *below,*
and a Splashing tetra (Copella arnoldi).

The two species of fish shown here are characoids; they possess certain anatomical internal features in common. Superficially, however, they look different, both from each other and from the majority of characoids. Like most characoids, *Copella arnoldi* has eyes. Unlike most, it lacks the adipose fin. *Anoptichthys jordani* has the adipose fin, which is present in most characoids, but lacks eyes and pigment.

Fishes have many ways of protecting their putative and actual offspring; bubble nests, mouth brooding, physical guardianship and so on. All of these methods, however, take place in water. As far as is known, *Copella arnoldi* is unique in being the only freshwater fish that lays its eggs out of water. This is a large advance for a modern fish, although admittedly, it was a step taken by archaic fish that were the ancestors of land-living vertebrates. An egg laid on land must have a waterproof cover, or shell. In reptiles this is leathery, in birds brittle, but both protect the egg from dessication. No bony fish has a shelled egg, all are gelatinous; hence the interest in *Copella arnoldi* laying its eggs out of water.

The site selected for the eggs is the underside of a leaf, two to four inches above the water surface. The male and female leap out of the water more or less simultaneously and the female lays sticky eggs on the leaf while the male fertilises them. From five to about a dozen eggs are laid at a time and the process is repeated until up to more than a hundred eggs are stuck on the leaf. When the spawn is laid, the male stays beneath the leaf and keeps the eggs moist by splashing water over them two or three times an hour. In two to three days the young hatch, drop into the water and swim away. The enlarged upper lobe of the caudal fin of the male is used for spraying the eggs, but it appears not to have been specifically evolved for that purpose. A similarly enlarged lobe is found in some closely related species which are known to deposit their eggs in pits in the river bottom.

Copella arnoldi comes from the lower Amazon, and it is in the New World that the greatest radiation of characoid fishes has occurred. The San Luis Potosi caves of the Yucatan Peninsula, Mexico are the home of the Blind cave characin.

Not all blind fish live in caves. They are found in the deep sea, living under rocks in rapids and even under stones on Californian shores. Nonetheless, as there are some 20,000 species of fish and only thirty-plus species of blind cave fish, the latter are of immense interest.

Cave fishes have been found in every continent except Europe, Asia west of the Urals, and Antarctica. Generally, it can be said that they lack eyes and are depigmented; the lack of pigment allows the blood flowing through the small capillaries to give a rosy hue to the fish.

A great deal of research has been carried out on a few species, one of which is the Blind characin shown here. The choice of this species as a subject for research is not because of its abundance in the wild but because it is the only cave fish so far bred in captivity. What has been learned is how poorly we understand cave fishes and their surface-living relatives.

Consider the Blind cave characin; it comes from the Yucatan peninsula where, living in the streams on the surface, is a very similar fish, *Astyanax fasciatus*. If this species were white and eyeless, it would be identical with *Anoptichthys jordani*. The absence of eyes and lack of pigment was considered sufficient justification to place these two forms in different genera. Cross-breeding experiments were performed and it was found that *Astyanax fasciatus* and *Anoptichthys* were fertile and the offspring of this match showed varying degrees of eye development. Even in the cave environment, the young of the Cave characin are born with a retina and a lens. These never develop and are eventually lost below the growth of skin. As the surface-living *Astyanax* is a lover of shade, we can begin to get an inkling of how this species of cave fish evolved. One of the problems that cave fish can present to the scientist interested in their relationships is whether the absence of eyes (and pigment in this case) is sufficient to justify separation at generic level from their epigean counterpart. In the case of *Anoptichthys* it would seem to be unjustified and the Blind cave characin may better be included in the genus *Astyanax*.

The great majority of cave fishes have close relatives living above them. But there are some existing exceptions. In the caves of Cuba, as well as of Yucatan, live the cave fishes *Lucifuga* and *Stygicola* (singularly apposite names). These two genera are brotulids, and almost all the other brotulids are deep-sea fish. The explanation for this odd situation is as follows: originally brotulids were secretive, shallow-water fishes, hiding in crevices and sea caves below the tide line. In the course of time, the land became raised, and some species were trapped in the caves (the process was, of course, extremely gradual). Slowly, as the sea water became replaced by fresh water, the fishes were able to adapt and survive in the new conditions. Meanwhile, their shallow-sea relatives had shunned light to a similar degree and moved into the deep sea, thereby enhancing the separation between two parts of previously contiguous populations.

In some cave systems, estimates of the number of individuals can be made, because the extent of the cave system is known and in a given area the number of fishes can be counted. In other cases, the number of individuals is unknown, for example an artesian well in Texas broke into a subterranean water supply and one specimen of a blind subterranean catfish was discovered. None has been seen since.

Plate 12

Three species of South American characin. In the foreground, a Piranha, above, the Rosy tetra (Hyphessobrycon rosaceus) *and left, a group of Neon tetra* (Hyphessobrycon innesi).

"This most voracious fish is found plentifully in all the rivers in Guiana, and is dreaded by every inhabitant or visitant of the river. Their jaws are so strong that they are able to bite off a man's finger or toe. They attack fish ten times their own weight and devour all but the head. They begin at the caudal fin and the fish being without its principal organ of motion is devoured with ease, several going to participate of the meal . . . Large alligators, which have been wounded on the tail, afford them a fine chance of satisfying their hunger, and even the toes of this formidable animal are not free from their attacks.

"Whilst we were continuing our course on the river Corentyn, one morning, an object was observed to drift into the middle of the stream around which there appeared to be a great commotion . . . When we came near, we observed the head of a large luganani or sunfish (*Cichla* spp), which was surrounded by numerous pirais, tearing off large parts of its flesh. We secured the luganani, which might have measured from twenty to twenty-six inches, and although the poor animal had been eaten off piecemeal to within its pectoral fins, it was still alive. Being deprived of its tail and pelvic fins it drifted perpendicular. The corial (a dugout canoe pointed at both ends) was brought to, our hooks and lines soon out, and we caught several of these depredators, which, with the remnant of the luganani, afforded us a good breakfast.

"The ducks and geese are equally exposed to the attacks of the pirai, and those which the settlers keep near the banks of the river are generally deprived of the lower part of their feet. It is a strange sight to see them walking about on mere stumps. In Wicki, a wood-cutting establishment on the river Berbice, there were two ducks which had been perfectly tamed by the Indians, and had been brought from the large ponds in the interior. Un-acquainted with the danger which the ravenous pirai offered them, their instinct directed them to their favourite element and one of them paid for its first visit with the loss of its toes and the other was similarly injured in its future visits. They now became cautious, and it was remarkable to observe how studiously they kept inshore, and never trusted them-selves beyond their depth."

The above, honest, account of the "Black Saw-Bellied Salmon", now the Black piranha, was written by the explorer Robert Schomburgk in 1841. Schomburgk, during his travels in South America (it was he who discovered the giant water-lily, *Victoria amazonica*) kept copious notes of his observations from which the above account was taken. It is rather regrettable that subsequent travellers have been much less honest and accurate and that the tales of piranhas' ferocity have become increasingly exaggerated. Piranhas are not large fish. They are deep-bodied, but rarely grow to more than about eighteen inches long. The jaw muscles are extremely powerful and the sharp-edged, triangular teeth can easily chop through a finger or remove cleanly about a cubic inch of flesh from a fish.

Fish are their main food, but they will willingly accept any other animal flesh that they can get. There is an area in the piranha legend where it becomes difficult to distinguish fact from fiction. Piranhas are shoaling fishes and a shoal can remove a large amount of flesh in a short time. The shoal is attracted by blood and the noises of disturbance in the water; a small group eating their prey are soon joined by another, much in the manner of sharks. It is within the bounds of possibility that a man with a bleeding wound could be attacked and killed, but whether this has happened as frequently as travellers' tales would have us believe cannot be corroborated. It is not impossible to believe the story of a man who fell into the water and was rapidly turned into a skeleton by piranhas whilst his clothing was left untouched, but it would be nice to have irrefutable proof of such events.

By contrast with the piranhas, the small brilliantly-coloured tetras are peaceable, attrac-tive denizens that justifiably have pride of place in many aquaria. It is difficult to define the term tetra, but it is loosely applied to small members of many genera including *Hyphes-sobrycon* (e.g. *H. innesi*, the Neon tetra and *H. rosaceus*, the Rosy tetra), *Cheirodon* (e.g. *C. axelrodi*, the Cardinal tetra) and *Hemigrammus* (e.g. *H. erythrozonus*, the Glowlight tetra).

Why do the tetras have such conspicuous coloration and markings? Do the bright colours make them more liable to predation? We can go some way to answering these questions at a general level, but not always at a specific one. Characins are mostly diurnal fishes with acute eyesight and many species live in small shoals. The advantages of these two features can be seen in the X-ray fish, *Pristella riddlei*, for the black spot on the dorsal fin is known to act as a shoaling stimulator. When a fish is alarmed, the dorsal fin is jerked rapidly, which causes the others to form a shoal and follow the fish giving the warning signals.

Brilliantly-coloured species like the Neon tetra and the Cardinal tetra live in rather dark, forest streams. It has been suggested that the shining colours help the fishes to find each other after they have been scattered by floods and so build up viable breeding populations again. Black markings seem to be commonest in mid-water shoaling species and the brilliant metallic colours are primarily used in close contact between individuals especially during courtship.

Characoids are found in both Africa and South America. They were originally inhabitants of the great southern continent Gondwanaland which split up over sixty million years ago to form the two continents either side of the South Atlantic. At the time of the Gondwana break-up, the cyprinids had not invaded as far as the proto-South-American portion, and therefore the characoids there have been able to evolve into more than a thousand species without competition from the cyprinids. In Africa, there are comparatively few species of characoids, none of which have the extremes of form or colour shown by their transatlantic relatives. It is thought that their speciation has either been hampered by the presence of the cyprinids or that only a very few ancestral species were left on the African part of Gondwanaland; the South American portion taking the lion's share.

New characins are frequently being discovered in South America, which shows how little we know about the fishes of the Green Continent and also how localised is the distribution of some species.

Plate 13

A spiny eel (Macrognathus aculeatus) *below, and two African killifish,* Roloffia occidentalis *centre, and* Aphyosemion geryi *above.*

The spiny eels are not true eels but the spiny part of their name is, for once, an accurate description. Spiny eels comprise the family Mastacembelidae with genera *Mastacembelus* and *Macrognathus*. They are advanced fishes found in Africa and parts of Asia. The spines referred to are found in a row along the back from behind the head to the dorsal fin. The spines are small, but can number over thirty, and as they point backwards (even when erect), it is difficult to know what value they have to the fish under normal circumstances. Under abnormal circumstances, their value is well known. If one tries to pick up a live *Mastacembelus*, say from a line or out of a net, the fish reacts by vigorously wriggling backwards so that its spines ram into the interfering hand. The usual result of this on the unwary is to drop the fish which will then escape.

Although mastacembelids are not eels, it is not at all clear what they are or to what other fishes they are related. True eels (like the common eel *Anguilla*, the Conger eels and the Moray eels) are characterised, amongst other features, by having a larval stage. This larval stage is a slender leaf-like transparent creature called a leptocephalus, which metamorphoses quite rapidly (after two or three years in some cases) into what is quite recognisably a baby eel. True eels also lack spines. Parenthetically, one should mention that the electric eel is neither a mastacembelid, nor a true eel, but is related to the characins. The spiny eels are secretive creatures living at the bottom of lakes and streams, frequently buried with only their heads poking out. They are nocturnal and emerge at night to feed on insects. The largest species are from South-East Asia where one species, *Mastacembelus armatus*, grows to nearly a metre long. The larger species are used as food, and, dried and/or smoked they are regularly sold in markets. The African species are much smaller and they are not valued much as food.

There are, however, two interesting facets to the African spiny eels. One is that there is a small species flock of *Mastacembelus* (seven species) in Lake Tanganyika, and one of these species, *Mastacembelus cunningtoni*, lives at depths of fifty metres.

The most remarkable spiny eel lives only in the rapids downstream of Kinshasha in the Zaire river. From Pool Molembo (formerly Stanley Pool) the great river narrows to less than a mile in width and begins its descent towards the sea. These rapids are the longest in Africa and amongst the most severe in the world. Fishes are usually able to exploit an environment and the rapids are no exception. A fish like a *Mastacembelus*, used to hiding in crevices and under stones, would stand a good chance of successfully colonising such an environment, and so it has happened. Within these rapids lives *Caecomastacembelus brichardi*, a blind white spiny eel that spends most of its time buried. This extremely modified species (which parallels the blind cave fishes), was, because of its eyelessness and lack of pigment, originally placed in a separate genus, but a recent re-evaluation of its classification has returned it to the *Mastacembelus*. The same rapids have a unique fauna,

with a blind white cichlid and several species of catfishes with very reduced eyes. The sheer physical difficulties of collecting specimens in this region prevent us from knowing the full extent of this remarkable fauna.

In complete contrast to the rigours of life in rapids we come to fishes living in pools, swamps and slow-flowing forest streams – the killifish. Killifishes (also called Top minnows, or Tooth carps) are cyprinodonts, but unlike the cyprinodonts on plates 15, 16 and 17, they are egg-layers. They are a widespread group, representatives being found in Africa, in some of the countries bordering the Mediterranean, the Near East, South America and the southern part of North America and South-East Asia, as well as on offshore islands of these regions. The prettiest genera, *Aphyosemion* and *Roloffia* (*Aphyosemion geryi* and *Roloffia occidentalis* exemplifying the genera) are native to West Africa, whilst the brilliantly-coloured *Nothobranchius* species are found in East and Central Africa.

Among the killifishes, some species show remarkable adaptability to and resilience in, difficult environments. *Oryzias latipes*, for example, is native to Japan, and there it has been seen swimming around under the ice in open ponds. In contrast, *Aphanius dispar* from between the Dead Sea and Iraq, has been found in hot springs at about 100°F (38°C) as well as in brackish to strongly saline water.

For many of the species inhabiting shallow, temporary pools in the tropics, drought presents a problem. The killifish have been able to ensure the continuation of their species even under these conditions. Some of the South American forms (e.g. genera *Cynolebias* and *Pterolebias*) and some African forms (*Nothobranchius* and *Aphyosemion*) lay drought-resistant eggs. Although the parents die (hence they are called annual fishes), the eggs survive the drought and hatch at the start of the next rainy season. The parents mate before the start of the dry season and the fertilised eggs are buried in the mud. The eggs continue to develop normally until the water dries up, when the embryo, protected by the dessication-proof egg membrane, becomes dormant. Shortly after the rains have started again, the eggs hatch and the cycle is repeated. An interesting fact is that, although in the wild the parents will inevitably die as the pool dries out, it seems that in captivity they will still die after spawning, even though their water is permanent. It also appears that for some species kept in captivity (although not all have been adequately studied), it is necessary for the eggs to pass through this dormant phase. Aquarists report that when the eggs have been so dried and replaced in water the embryos all hatch at almost the same time, within limits, irrespective of the original time of laying.

Certainly, the degree of adaptation is profound, but it must be an important part of the life cycle of some species because there are some that have only ever been found in pools that regularly dry out each year.

Plate 14

In the centre, a pair of wild Gambusia affinis *(the male below).*
At the corners, four examples of cultivated male guppies (Poecilia reticulata).

The often-used expression "live-bearing fishes" is not particularly instructive; for if a species is to survive, its young must be born alive and not dead. The same expression also covers some rather different biological processes. Mammals give birth to fully formed young. They are "live-bearers"; by this it is meant that the very small egg is fertilised inside the mother and that the growing embryo, not having a yolk sac for nourishment is nourished from the food in the mother's blood. The food being made available via the medium of a placenta, this process is true viviparity. In the world of fishes, some sharks have an analogous process, even to the extent of a pseudoplacenta being formed. A few bony fishes, such as the Four-eyed fish (*Anableps*) have a form of "placental" nourishment. In the case of *Anableps*, the young develop in follicles in the ovary and the follicular walls are specially adapted for gaseous and nutritional interchange but the embryo-maternal contact is only superficial – there is no equivalent of the umbilical cord.

The majority of "live-bearing" fishes, especially within the Poeciliinae (the tribe most familiar to aquarists, i.e. the guppy, mollies, platies, swordtails, etc.) are ovo-viviparous. By this is meant a state in which, to varying degrees, the eggs develop and hatch inside the mother, rather than in the water. The object of this adaptation is to try and eliminate the period when predation is greatest upon the species, the free egg stage. The developed, free-swimming young stand a much greater chance of survival after birth than does a newly extruded egg. In some species, the young fish relies largely on its yolk sac to provide food when within its mother; in other species, trends leading in the direction of pseudoplacental development are evident. Even within one or two of the egg-laying cyprinodonts there are signs of incipient ovo-viviparity. The female Japanese medaka for instance, may keep the fertilised eggs within her body for a little while before laying them. Within the cyprinodonts then, one can observe all stages from straightforward oviparity to the development of pseudoplacentae.

In mammals, which have a constant body temperature, the period of gestation is predictable because the rate of development of the embryo is constant. The internal temperature of fishes varies according to the water temperature so that the gestation period is variable; in warm water it is much shorter than in cold water. Again, unlike mammals, there is a rough correlation between the size of the mother and the number of young. A small female guppy will only produce a few young, whereas a large female can drop over sixty.

In egg-laying fishes, fertilisation of the eggs occurs when the male scatters his sperm (milt) in the water near the eggs. Such a mechanism would scarcely be adequate for the ovo-viviparous fishes, and so internal fertilisation has been developed. Naturally this involves certain morphological changes in the male in order that the sperm is directed to where it is needed. Male poeciliines have developed a structure called a gonopodium, formed from elongated rays of the anal fin. Depending upon the species, the gonopodium is formed essentially of the third, fourth or fifth ray of the anal fin. The ends of the elongated rays may be hooked or spatulate and are used in the transference of sperm. The shape of the gonopodium and the nature of the distal expansion, the presence or absence of hooks etc., are specific features and important in identifying species. Concomitant with the enlargement of anal fin rays comes the development of specialised muscles for moving the gonopodium. The female's anal fin remains unmodified and retains (in the case of the guppy) the fan-like shape present in the juveniles before sexual differentiation starts.

There are other, more obvious differences between the sexes. The males are smaller and much more brightly coloured. It is the variable bright colours of the males that has led to much selective breeding to produce forms more exotic than were ever found in the wild.

The fish in the centre of the painting is *Gambusia affinis*, a species native to the southern and eastern states of the USA. The males of this species are less spectacularly coloured than the male guppy, but have many pleasing iridescent reflections. Black and yellow individuals are known from the wild. This seemingly inconspicuous fish is now found in many tropical fresh waters where it has been introduced for mosquito control. Like many cyprinodonts, it is a surface feeder and eats the egg rafts of mosquitoes as well as the larvae and pupae. Of this fish, the great American aquarist, William Innes, has written "Success in building and maintaining the Panama Canal depended partly on the solution of the fever problem. *G. affinis* was, and still is, responsible for making Panama inhabitable to the white man." *Gambusia affinis* has been introduced into Africa, the Near East, Malaysia, China and Japan, often, regrettably, to the detriment of the local fish populations. It is a singularly hardy species, being able to tolerate brackish water and a temperature range from 12° to 35°C. At the higher end of its temperature range, a greater proportion of black (melanistic) individuals occurs.

The guppy is native to South America north of the Amazon, Barbados and Trinidad, but, like *Gambusia affinis*, has been introduced into many parts of the world to control mosquito populations. The common name commemorates the Rev. Robert John Lechmere Guppy who brought the first specimens of *Poecilia reticulata* from Trinidad.

The separate, wild populations of guppies tend to have different, yet characteristic dominant colours and colour patterns. The wild populations in Florida, for example, tend to be copper-coloured with black spots on the tail. Wild females, however, are a uniform olive-brown with perhaps a few dark spots. Selective breeding and cultivation, which has been practiced since just after the First World War has not only produced the exotic males (examples of only a few of the recognised type are shown here), but has also been able to produce females with red, black or iridescent colouring on the body (usually the posterior half) and with marked and coloured fins.

The guppy has a gestation period of four to six weeks, and, as a female can produce over sixty young which become sexually mature in two months, the breeders have a comparatively easy time selecting for, and improving, particular strains.

Plate 15

A pair of wild green swordtails (Xiphophorus helleri) *bottom right, and, facing them, a pair of red swordtails especially bred from the wild form. Bottom left and top right, two varieties of the platy,* Xiphophorus maculatus.

The common name platy derives from the formerly-used generic name *Platypoecilus*. Now, however, the platy is considered to belong in the same genus as the swordtail and so becomes *Xiphophorus maculatus*. Although *Xiphorphorus* means sword-bearing, and the platy never develops a sword, swordtails and platies are very closely related and can produce fertile hybrids when bred together in an aquarium.

All the species in the genus *Xiphophorus* are confined to the eastward-flowing rivers of Central America from about 26°N to 15°N. The platy is native to the southern part of Mexico, Guatamala and northern Belize. It is a singularly variable species. Quite different, but consistent and recognisable colour patterns occur not only from river to river, but also in different parts of the same river. In fact, several writers have independently described the platy as the most variable vertebrate.

A consensus might describe the basic (or most frequently encountered) wild form as having a body colour from yellow or olive-brown to blue-grey, usually with a pattern of dark spots on the posterior part of the body. The gill cover is a metallic blue-green. Males frequently possess a few faint transverse bars on the body, females very rarely do.

Xiphophorus maculatus is usually found in the quieter, gently-flowing, lower reaches of rivers, and in lowland ponds where the annual temperature range is not very wide and is also relatively low.

The natural variation displayed by the platy (or moonfish as it is also known) has been cultivated by aquarists to produce the great number of differently shaped and coloured varieties now available.

Compared to the platy, the swordtail (*Xiphophorus helleri*) might be said to represent the other extreme of the genus in that the lower part of the tail fin of males is prolonged into the "sword". Incidentally the sword referred to in "*Xiphophorus*" is probably the gonopodium and not the elongated lower rays of the tail fin. Other species within the genus show varying degrees of "sword" (*sensu* tail) development. The sword is absent in the platy and *Xiphophorus couchianus*; just detectable in *Xiphophorus variatus*; small but clearly present in *Xiphophorus xiphidium*; larger in *Xiphophorus montezumae* and most developed in *Xiphophorus helleri*.

Like the platy, the swordtail is a variable species. The body shape and colour of the swordtail is affected by two factors, the genetic make-up of the population and the environment. For example there are four separate subspecies, but, at the same time, individuals or populations of each or any of these subspecies can vary. Fishes living in the faster-flowing head waters of a river will have a thinner body, a shorter sword, a smaller dorsal fin and frequently, smaller scales. Even the gonopodium varies with the environment, that of males from faster-flowing waters has enlarged hooks to help in securing the male to the female during mating. The wild form is represented in this painting by the Green swordtail and the variations in bodily proportions, relative length of sword, colour, presence or absence of vertical bars on the body and the partial duplication and colour of the red

lateral stripe, are used to define the various subspecies and local populations. Nonetheless the wild form is the Green swordtail and the selective breeding of individuals showing particular characteristics, and crosses with other species of the genus, have produced the many other varieties known.

Black is not a conspicuous feature of wild *Xiphophorus helleri*, yet various varieties of swordtails frequently seen in aquaria have either large black blotches or a completely black body. The black colour (in the case of the Hamburg swordtail) was produced by crossing a green female *Xiphophorus helleri* with a red male *Xiphophorus maculatus* (platy). The red male platy had, however, first to be produced by selective breeding from wild forms showing a preponderance of red.

Although the various species and subspecies of *Xiphophorus* will interbreed, not all crosses are either productive or successful. The production of offspring with a lot of black pigment (melanin) is a case in point. Certain of the genes in the platy that carry the instructions for the production of melanin can, if introduced in *Xiphophorus helleri*, cause the manufacture of so much melanin that skin tumours result. The same genes, if introduced into *Xiphophorus xiphidium*, will cause the production of the desired black pigment but will not cause the skin tumours. There are however other black-carrying genes in *Xiphophorus maculatus* which will not produce skin tumours in a cross with *Xiphophorus helleri*. A working knowledge of genetics is an advantage to a fish breeder!

If this brief illustration of the potential existing in the Green swordtail and the wild platy that can be released by careful breeding in aquaria is not an adequate demonstration of the unusual qualities of these species, then consider sex reversal.

Not infrequently, a large female swordtail has produced, over a year or so, many dozen young for the benefit of the breeder when she ceases so to do. The dark "pregnancy marks" between the anal fin and pelvic fins disappear and slowly but surely the lower rays of the tail fin begin to elongate. While the sword is developing, the anterior rays of the anal fin also elongate. The body becomes thinner, the sword develops fully and the anal fin rays form the gonopodium. The ovaries have regressed, testes have developed and become functional so that the ex-female is now a fully functional male, capable of fathering young. The change from female to male is extremely common, but the reverse change has not been observed.

The sex of humans is determined at the moment of conception, but the situation is rather different in swordtails. The young and adolescent fish are potentially of either sex. If the ovaries develop first, the hormones they secrete suppress the growth of the testes (and therefore the male secondary sexual characters) and so the fish is a functional female. If, later on, the ovaries cease to function, the male organs are no longer suppressed so they develop and function, hence the fish effectively and practically changes sex. A male-to-female change would be more difficult from the point of view of the secondary sexual characters which would have to change from the specialised male condition back to the generalised female condition.

Plate 16

A *Giant sailfin molly* (Poecilia velifera) *left, and a Marbled molly* (Poecilia latipinna).

Like the platies on the previous plate, the mollies derive their common name from the formerly used generic name *Molliensia* which was given to these fishes by Le Sueur in 1821 in honour of the then French Minister of Finance. The nine species of molly are now included in the genus *Poecilia* along with the guppy. Not all the mollies are black, nor do the males of all species have the large sail-like dorsal fin of the species portrayed here.

As one would expect by now, the mollies have many of the characteristics typical of the poeciliines described in connection with the previous two plates. Two attributes typical of live-bearing fish, shown by the mollies but not previously mentioned, are sperm storage and super foetation.

The females of some species have the ability to store sperm in the lining of the ovary and keep the sperm alive for up to a year. During this time, the female may produce several broods of young, all the eggs having been fertilised by the initial deposit in the "sperm bank". The details of this process are not yet fully understood. Along with the storage of sperm can come the phenomenon of superfoetation. This means that at any given time, there are several broods within the mother at different stages of development. One of the Mosquito fish (*Heterandria formosa* – which, incidentally is one of the smallest vertebrates, a fully-grown male being only three-quarters of an inch long) has been observed to drop young every few days, thereby reflecting the rate at which the eggs are fertilised.

The live-bearing poeciliines are entirely New World in origin, and the distribution of the tribe is almost the same as that of the genus *Poecilia*, which, with over thirty species, is the second largest genus of the tribe. They occur on the coastal plains of the USA from Virginia in the East, through East and West Mexico, Central America, the eastern side of South America down to La Plata and at various sites within the Amazon basin. The great majority of live-bearing poeciliines live north of the equator.

There is a great deal of interest centred on the Amazon molly (*Poecilia formosa*). In 1932, Professor and Mrs Carl Hubbs, a noted American husband-and-wife ichthyological team, discovered that wild populations consisted only of females; no male had ever been found. Males of two other species, *Poecilia sphenops* and *Poecilia latipinna*, occurred in the same waters along with their respective females, but of males of *Poecilia formosa* there was no sign. The Hubbs concluded that they had found an example of gynogenesis. The sperm of the males of the quite distinct other species were used only to stimulate egg development and did not contribute any genetic material to the offspring.

Maleness in live-bearing fishes is determined by the odd chromosome present only in the sperm. If, therefore, the sperm does not contribute its chromosomal complement to the egg but merely acts as a mechanical stimulation to initiate cell division then, by definition, the female can only produce female offspring.

A sample of over a thousand Amazon mollies collected in 1958 contained one male that was neither a *Poecilia sphenops* nor a *Poecilia latipinna*, and which was considered to be a true male *Poecilia formosa*. The true identity of this lone male cannot satisfactorily be determined on the basis of one dead specimen. If more of these males could be collected alive and shown to breed with female Amazon mollies and produce young of both sexes, then its identity as a male *Poecilia formosa* is established beyond all reasonable doubt. If the offspring were all female, then the male in question is not a male *Poecilia formosa* but a mutant of another species. The suggestion has been made that the Amazon molly, although widespread on the Atlantic coast of Mexico and Panama, may in reality be a natural hybrid. A guess as to which species were the parents of the hybrid has not yet been hazarded. Parenthetically, it should be noted that a similar, but more confusing, situation exists in another poeciliid, *Poecilopsis*. A population has been found in northwestern Mexico in which there are two types of female. Although identical to look at, one type produces only female offspring, while the young of the other type has a normal sex ratio.

Three species of molly have very large dorsal fins when adult; *Poecilia velifera*, *Poecilia latipinna* (shown here) and *Poecilia petensis*. The last species is rarely imported, but it is of interest in that the lower rays of the caudal fin of the male are produced into a small spike, thus paralleling the situation in the swordtails.

The Giant sailfin molly, *Poecilia velifera*, is one of the few poeciliines to live in brackish water. Its natural habitat is the coastal region of southeastern Mexico where it lives in streams and lagoons where the salinity may reach fifteen percent of that of seawater.

The Marbled molly is a cultivated, or at least stabilised, variety of *Poecilia latipinna*. This species has the most northerly distribution of any in the genus. It is found along the eastern seaboard of the USA from Virginia to the Rio Grande on the Mexican border. Even in the wild, this species has produced considerable differences in coloration. Individuals from almost albino to completely black have been found. The common "Black molly", obtainable for a few pence in most pet shops was especially bred by a breeder in New Orleans. It took him seven years to perfect the uniform black colour with its peculiar velvety texture. The Black molly is allegedly a hybrid between *Poecilia sphenops* and another species. The specific identity of the other parent is unknown. The original breeder cited *Poecilia formosa* as the female parent but if this were so, then all Black mollies would be female. If the origin of the original Black molly is in doubt, the identity of many Black mollies now on sale is even more uncertain. *Poecilia sphenops* can certainly be induced to produce various long-finned or lyre-tailed varieties, but cannot, on its own, increase the number of rays in the dorsal fin to the number found in many of the current Black mollies. Certainly from the presence of characters not existing in *Poecilia sphenops* but present in the Black molly, other species, wittingly or unwittingly have been involved in the production of the current common Black molly.

Plate 17

The Firemouth cichlid (Cichlasoma meeki) *left, and the Zebra,*
or Convict cichlid (Cichlasoma nigrofasciata) *right.*

This painting is the first of three concerned with a family of advanced "perch-like" fishes very popular with aquarists – the cichlids. Almost all the cichlids are native to Africa and South America: the few exceptions are two species of the genus *Etroplus*, which are endemic to southern India and Sri Lanka; some species from Syria, Israel and Jordan; some species in the southern part of the USA; and the cichlids living on Madagascar.

The species in the Near East are thought to have moved northwards from Africa to the Jordan valley before the climatic changes which produced the deserts. After the emergence of the land connections between North and South America, some species migrated northwards from South America as far as Texas. The Madagascar forms are believed to be secondarily freshwater species and to have evolved from salt-tolerant species that were able to cross the Mozambique channel.

The distribution of the cichlids, like the distribution of the characoids, is of great significance in the development of the theory of continental drift. South America and Africa were once part of a great southern continent, Gondwanaland, which broke up about sixty million years ago. Therefore, any freshwater fish groups common to the continents on either side of the south Atlantic must have been present before the break-up. India and Madagascar were also, as was Antarctica, part of Gondwanaland, but, as suggested above, one must exercise a certain amount of caution with regard to the presence of cichlids in Madagascar. The relationship of the Indian cichlids to the African ones is not known. They could be primitively Indian, or they could have island- or coast-hopped but left no traces of their passage. Their evolution, after their arrival in India, need not have followed the same evolutionary path as their closest relatives left behind in Africa. Such zoogeographical speculations are stimulating armchair zoology, but not necessarily conclusive.

What characterises a cichlid and why are there so many species of these fish? The answer to the first question is easy. Cichlids have only one nostril on each side of the head. Almost all other fish have two, one to admit the water current to the olfactory surface, the other to let the water emerge – a one-way current. Cichlids achieve the same ends by an in-out method. Another feature of cichlids is the double lateral line. The longitudinal row of pore-bearing scales (the lateral line), is the external manifestation of a sensory system; the pores are the openings which allow disturbances in the water to be recorded by sensitive devices in the canal joining the pores below the skin. In most fishes there is one row of pores along the middle of each side of the body. In the cichlids (and in some of the labyrinth fishes) there are two rows: one row starts in front of the tail and extends part of the way towards the head, the other starts behind the head and extends part of the way towards the tail, higher up than the more posterior line of pores. Another important feature of cichlids, and one of great diagnostic importance, is the pharyngeal bone. This is an approximately triangular tooth-bearing bone lying on the floor of the "throat". Its function is to further break up food, which it does by rubbing against a hard pad at the base of the skull.

Why are there so many species of cichlids? A complete answer to this question is not available, but various contributory factors can be identified. Firstly, the construction and mechanics of the body, especially of the head and jaws, is such that changes to utilise different foods can be relatively easily and successfully effected. Secondly, the behaviour of cichlids lends itself to speciation. The courtship rituals and the colour pattern of the males and females are precise, and any adventitious mutations in colour or behaviour may, in the wild, limit the choice of potential mates. A reduction in the choice of available partners will limit the amount of interchange of genetic material and so genetic incompatibility between the parent form and the mutant form could result. Another behavioural factor is territoriality.

A pair, or one of a pair, will select a site for breeding; and a corollary of this territoriality is parental care. Care of both the eggs and young ensure the highest possible chance of survival for the young. The care can take the form of incubating the eggs and protecting the young in the mouth of one of the parents or, in the case of the Firemouth cichlid (*Cichlasoma meeki*), the eggs, once laid, are guarded by both parents. When the eggs hatch, the young are kept together in a shoal, and in the case of danger are herded by one parent while the other attacks the source of danger. It has been reported that several species feed their young by chewing up large food particles, masticating them with the aid of the pharyngeal bone and then spitting out the fine food particles over the shoal of young. *Cichlasoma cutteri* is notable among cichlids because its eggs have stalks which adhere to the leaves of plants. An extreme case of parental care is shown by the deep-bodied South American Discus fish (*Symphysodon* species). Secretions emanating from the skin of the parents attracts the young. These secretions also serve as food for the young in the earliest stages. An interesting observation, which has often been made, is that some species of cichlids will pair for life, although this has not been corroborated satisfactorily and may not necessarily aid the success of a species.

Generally speaking, the South American cichlids have developed more deep-bodied forms than have the African cichlids. The Angel fish and the Discus fish exemplify this. There has also been a radiation of dwarf forms in South America (*Apistogramma* species) which has no real parallel in Africa.

The Zebra, or Convict cichlid (*Cichlasoma nigrofasciata*) and the Firemouth cichlid are both inhabitants of Central America. The common names of both species do not require any explanation. The expansion of the red membranes below the gills of the Firemouth cichlid serves as a warning to would-be adversaries. The function of the red inside of the mouth is not certain, although it has been suggested that it is used in concentrating the shoal of young fishes.

The sense of territory, coupled with the feeding and habitat specialisations, could, in a once-widespread species, increase the number of species with time. For instance, imagine a rock-dwelling species living round the edges of a lake. If the water level changed in such a way as to produce patches of open water between rock outcrops, the fish may be deterred from moving from their own outcrop to another. The populations on each outcrop would breed only amongst themselves and in time may become genetically isolated. It has been suggested that the large species flocks of Mbuna in Lake Malawi may have originated in this way.

Plate 18

The Nigerian mouthbrooder, a fish known to aquarists as Haplochromis burtoni.

The cichlids are a group of fishes which have long held a unique fascination. In the early days, the Egyptians valued *Tilapia* species for food, and painted accurate pictures of them on the walls of their tombs. Today, although *Tilapia* are still extensively cultivated for food in many parts of Africa (and have been introduced into tropical Asia), the main interest in these fish lies jointly in the world of the aquarists and the world of the scientists. The former like cichlids because of their bright colours, the changeability of the colours, their behaviour and the high chance of breeding them in captivity. The latter group are interested in cichlids because of their distribution, morphology, behaviour, speciation and evolution.

Considering all the problems of evolution with which cichlids puzzle scientists it is peculiarly appropriate that Tom Adams should have chosen to paint the fish that he did. No one is certain what species it is. Although called the Nigerian mouthbrooder and sold (usually) under the name of *Haplochromis burtoni*, there is no certainty that the species is indigenous to Nigeria, and, equally it is not necessarily *Haplochromis burtoni*, which comes from Lakes Malawi and Tanganyika.

There are two major lineages within the African cichlids – the *Haplochromis* lineage and the *Tilapia* lineage (although it should be noted that many species of *Tilapia* have now been placed in the genus *Sarotherodon*). Many African cichlids can be confidently placed in one or other of these lineages, which are categorised by the particular arrangement of bones at the base of the skull. Currently, much scientific research is directed towards gaining an understanding of the *Haplochromis* species which are remarkable for their "species flocks". There are very few species of *Haplochromis* in rivers; the few river-dwellers which do exist are moderately widely distributed and are generalised; i.e. they do not display the remarkable specialisations shown by their congeners. Yet in Lake Victoria there are well over one hundred and seventy species of *Haplochromis*, and all save some half-dozen live only in that lake, and evolved therein. Similar statements are equally true of the other large lakes, Malawi and Tanganyika, although because the evolution is more advanced there (these lakes are much older than Lake Victoria), the species have diverged further (in morphological terms) and are now placed in many different genera. Even so, the lineage to which they belong can usually be determined. It is these accumulations of species which have evolved in the great lakes, that are called the species flocks. There are a few examples of much smaller species flocks e.g. a few species of *Cichlasoma* endemic to Lake Nicaragua and a flock of less than half a dozen *Tilapia* species in Lake Malawi in the isolated Barombimbo in Cameroun. To be accurate, it must be pointed out that, although a great number of the one-hundred-and-twenty-plus cichlid species in Lake Tangyanika are haplochromines, it is not certain how many, or even if the remaining species belong to the other major lineage or if they have descended from several ancestral forms. More research needs to be done before any definitive statements can be applied to that part of the Lake Tanganyika species flocks.

Time is an important factor in the degree by which the species differ from each other: the earlier they diverged, the greater the differences. This leads to an interesting observation which has enabled some measurement of the rate of evolution to be made. Lake Nabugabo is a small lake, some twenty kilometres long by eight wide, situated at the north-western edge of Lake Victoria and separated from it by a series of sandbars. Due to the various changes in the earth's crust in this part of Africa, Lake Nabugabo is some fifteen metres above the level of Lake Victoria. We do not know exactly when the sandbars developed which separated the two lakes, but in the sandbars on an old shore-line of Lake Victoria, the same height as Nabugabo, some charcoal was found which was dated by radio carbon to about four thousand years ago. This must give a reasonable approximation to the age of Nabugabo. Lake Nabugabo has five species of *Haplochromis* that are endemic, so these five species have evolved in the last four thousand years, which in geological terms, is no time at all. Each of these five species has a close relative alive in Lake Victoria today.

It is an interesting speculation that, if Lake Nabugabo were to be reconnected with Lake Victoria, the number of cichlid species would immediately be increased by five. And it has been suggested that this process may well have occurred in the past. Certainly, in the rift valley of Africa, earth movements are far from rare.

The most conspicuous difference between the Nabugabo cichlids and their Lake Victoria relatives is that, in at least four cases, the males have a different breeding dress, a factor which, allowing for the precision in courtship shown by the cichlids, will tend to enhance the separation of closely-related forms. Even if the closely-related forms could produce fertile young, their breeding may be naturally discouraged, because the appropriate responses may not be elicited.

As has already been stated, the few species of riverine *Haplochromis* which do exist are generalised. They show none of the specialised adaptations for utilising particular foods, nor for living under a particular regime. They are smallish and although predominantly carnivorous will eat almost anything: from riverine fishes like these, the remarkably diverse species of the flocks have evolved in the lakes. From the carnivorous or omnivorous forms have evolved species with flat teeth and strong jaws which specialise in crushing molluscs; large pelagic species with big mouths that are piscivorous; species with chisel-like teeth that graze on algae and species with tweezer-like front teeth that pick up small insects and arthropods from cracks in the rocks. There are species with a prognathous lower jaw armed with small rasping teeth that feed on the scales of other fishes, and species adapted to feeding on the eggs or young of other species (how they induce the mouth-brooding parent to release the young is not certain, but it has been suggested that the distensible mouth of the predator engulfs the head of the brooding female and under risk of suffocation, she lets go the young). Most bizarre of all perhaps is *Haplochromis compressiceps*, a fish with a long, compressed head, narrow pointed teeth and a protruding lower jaw which eats the eyes of other fishes.

Plate 19

The South American Angel fish, Pterophyllum scalare, and a recently developed black strain.

The Angel fish shown here are cichlids of the genus *Pterophyllum* from the warm rivers of South America. One of the difficulties caused by giving common names to fishes is that unrelated species may receive the same name. The term "Angel fish" is used in America for members of a family of Coral reef fishes – the Chaetodontidae (called Butterfly fishes in England) whilst another family of Coral reef fishes – the Pomacanthidae – are called Angel fishes in England. "Angel fish" is also used as an alternative name for the monkfish, a bottom-dwelling marine fish related to the sharks and rays; this Angel fish (*Rhina squatina*) is not particularly pretty (although beauty is a subjective concept). It grows to about six feet long and weighs up to seventy pounds.

The first Angel fish (*Pterophyllum*) brought back to Europe were collected in Brazil by the explorer-naturalists, Spix and Von Martius, who journeyed throughout that country from 1817 to 1820. Their specimens were described by Lichtenstein in 1823 as *Zeus scalaris* and are now known as *Pterophyllum scalare*. Castelnau described *Plataxoides dumerilii* (now *Pterophyllum dumerilii*) in 1855 from the Amazon near Belem. *Pterophyllum altum* was found in the Upper Orinoco and named by Jacques Pellegrin in 1903. The species thought by many to be the commonest species in captivity, *Pterophyllum eimekei*, was not recognised until 1929 and then, in 1963, J. P. Gosse described yet another species from the Rio Solimoes which he called *Plataxoides leopoldi*. The confusion over the usage of the common name Angel fish is present again in the scientific names. It is now generally thought that *Plataxoides leopoldi* is the same species as *Pterophyllum dumerilii*, which left us with four species. Leonard P. Schultz reviewed the Angel fishes in 1967 and came to the conclusion that *Pterophyllum eimekei* is really *Pterophyllum scalare*. He argued that the description *Pterophyllum eimekei* included counts and measurements from inbred aquarium specimens and that the slight difference described between "*P. eimekei*" and *P. scalare* do not exist in wild populations. Schultz, therefore considered *Pterophyllum eimekei* to be a synonym of *P. scalare* and that there are only three species of Angel fish. *P. altum* (Pellegrin), *P. dumerilii* (Castelnau) and *P. scalare* (Lichtenstein). Either this work has not been universally accepted, or subsequent authors of books on aquarium fishes are unaware of it, for the names "*Pterophyllum eimekei*" and "*Pterophyllum scalare*" both still appear, often accompanied by an account of the colour differences between them. There is, however, a rider often added to the effect that the new aquarist may experience considerable difficulty in distinguishing these species. Fortunately for the readers (and the author) of this book, the fish that Tom Adams has painted purports to be *Pterophyllum scalare*.

The three species of *Pterophyllum* are very similar. The body is extremely thin in front view and almost circular in side view. The middle rays of the dorsal fin and the anterior rays of the anal fin are greatly elongated and make the fish look deeper than long. The few rayed pelvic fins are elongated into streamers which can reach back beyond the tail fin. With its black vertical stripes on an olive-silvery body, the fish may look very conspicuous in an aquarium, but in its natural surroundings it will be inconspicuous. The stripes are analogous to those of the zebra; they serve to break up the body's outline, so that, when hiding among weeds in slow-flowing or still waters and gently waving its long fins, the Angel fish would be very hard to spot from the side. From the front, the fish is so thin that when it is aligned parallel with fronds of water plants it would also be well hidden. Its secretive habits would seem to be for its own protection against predation, rather than to save alarming its prey. In the wild, *Pterophyllum scalare* eats mostly zooplankton and plants.

The Angel fish would seem to be the essence of silence as it gracefully cruises around the waters; in reality the male can be quite noisy. Normally they are territorial only during the courtship and rearing phases, but they may scrap at any time, and, as a precursor to aggression, produce clicking sounds with their jaws.

Pterophyllum (literally leaf-finned) species have been popular with aquarists for a long time. They were first imported alive into Germany in 1911 and ten years later into the USA. Spawning in captivity is not difficult and it was first achieved by a pair under the care of Mr W. Paullin in the USA in about 1930.

The eggs are laid on plants or other clean mid-water surfaces. Both parents guard the eggs until they hatch and then the young until they are capable of survival. When the parents are guarding the young, the latter are discouraged from straying by a change in the colour of the parents' pelvic fins. These become much paler in colour and the young seem to be conditioned to home in on the light colour.

It appears to be a foible of human nature to like deviations from the norm and the Angel fish has not escaped. From the wild-type Angel fish, many new varieties have been produced, hardly any of which could survive in the wild; indeed their absence in the wild suggests that they have already been selected against. Nonetheless Black (shown here) Black lace, Golden, Blusher and Long-finned angels can be bought at almost any petshop. All these varieties have been produced since that first spawning in captivity some forty years ago. It will be most interesting to see if, with several hundred years of selective breeding, the genetic complement of the elegant Angel fish will be capable of producing as many varieties as there now are of goldfish.

Plate 20

Siamese Fighting fishes (Betta splendens). *The wild form, top left, and three cultivated males.*

The three male Siamese Fighting fish (*Betta splendens*) shown in the centre of the painting represent three of the colour varieties produced by selective breeding. Selective breeding is a process by which certain characteristics potentially present in the wild form are allowed, and encouraged, to develop in captivity. Individuals showing the desired qualities are then bred with each other. For example, in the ditches and ponds of South-East Asia, the bright colours and the long fins would be selected against. The long fins would fray and the bright colours would be easily seen by predators. Only in the benevolent world of the aquarium are the fins and colours capable of flourishing and being inherited and exaggerated by successive generations.

In the wild, *Betta splendens* is an inconspicuous fish, the body is dark brown, often with a dark red hue and a dark horizontal line along the flanks. From this, bright red, bright blue, cornflower blue, pale yellow and even black fishes have been produced with fins which may or may not, according to choice of breeder and fancier, be the same colour as the body.

The shape and colour of the wild and domesticated forms may be very different, but the behaviour is little changed. The form of the bubble nest, the courtship ritual and the parental care are more or less constant and the species' most celebrated quality, its bellicosity, is undiminished.

The natives of Thailand have long appreciated the fighting qualities of male Fighting fish and for hundreds of years have wagered large sums of money, even fortunes, on the outcome of a battle between two males. It is not known when the Thais first recognised the fighting qualities of male *Betta splendens*, but it was a long time ago. Until about the middle of the last century, all the Fighting fish used in sporting contests were caught wild, but by then they had become scarce, or at least a regular supply within an easy travelling distance of centres of population could not be ensured. The answer was to keep and breed the fish in captivity. At first, the fish were kept solely for fighting, but later on the appearance of the fish became an increasingly important consideration in breeding.

In the early part of this century, the first coloured variety bred in captivity, a fish with a creamy red body and red fins, came into general circulation in Thailand. The Thai name for this variety was pla kat khmer (Cambodian biting fish) presumably because it first appeared in the stocks held by breeders in Cambodia.

The spectators enjoyment of these gladiatorial combats is not merely the thrill of betting, but watching the males' changing colours and postures as they threaten each other. In a local ditch, unthronged by spectators, such disputes between the fishes will not usually be fatal; the vanquished has room to move away from the threat postures and aggression of the victor. In a glass arena, however, death may result unless the fishes are separated.

Wild fishes kept in captivity will usually give up the fight after a few minutes, a bout lasting twenty minutes would be exceptional. However, selective breeding, choosing parents with a proven disposition for tenacious fighting, has produced fishes capable of fighting for hours. An unusually well-matched pair may scrap for twelve hours.

In a serious contest, the chosen combatants are usually of equal size. They are placed in adjacent glass jars so that they can see each other and threat postures are adopted. On the basis of the postures, bets are laid. The fish are placed together in a large vessel which gives the gladiators room to manoeuvre and gives the spectators a good view. A preliminary threat display occurs, fins and gill membranes are spread and colours flash across the body. The fighting starts when the fish line up, pointing in the same direction but with one fish slightly behind the other. This position is held motionless for a few seconds, then, suddenly, the fishes attack with a rain of bites so rapid that the eye can hardly follow them. Frayed fins, especially the unpaired fins are the first signs of injury. Fins and scales are usual, and sensible, points of attack and with two determined contestants the fins may finish up as frayed stumps. Sometimes there is a head-on attack where the jaws are locked as the two fishes roll around in a trial of strength.

The fight falls naturally into rounds; the Fighting fish breathes atmospheric air which is absorbed by a special organ at the top of each gill cavity. Even in well-oxygenated water, the gills cannot supply enough oxygen to the system and in a fight the oxygen deficiency rapidly builds up. Each fish therefore stops at will and goes to the surface to breathe and it is reliably reported that no sneak attacks occur during the breathing spells.

Only the minority of fights end in death; when one fish is too weak to continue, it will swim away when its antagonist prepares for another attack. The owners then separate the fighters. The deaths that do occur are either as a result of a secondary infection resulting from wounds or from shock and a loss of body tissue.

Astoundingly, the little male fighter is an excellent and gentle father. He makes a floating nest of bubbles stuck together by a sticky secretion. A gravid female is cajoled below the nest. The male wraps his body around hers until the eggs are extruded. This happens several times, for the first few matings are usually unproductive. As the eggs fall down, the male picks them up and spits them into the nest. There, the eggs are held by a combination of capillary action and the stickiness of the bubbles. Eventually the male has placed up to three hundred eggs in the nest and he stays below it, gathering any eggs that fall out. In three to four days the eggs hatch and the young shelter beneath the nest. Should they stray too early, the father herds the stragglers back to safety. If the mother has not travelled far enough away, she is chased off by the male to prevent her from eating any of the young. During this period, the male eats normally, but the eggs and young are never eaten. In about three weeks, the male leaves the young to follow their own lives. In a mere two months the young males start to show aggressive behaviour and the cycle starts again.

Plate 21

A male Lace gourami (Trichogaster leeri) and his nest.

Trichogaster leeri, the Lace gourami is an anabantoid fish. The anabantoids are a group of spiny-finned, old-world species that possess both functional gills and an accessory breathing organ. The accessory breathing organ (also called the labyrinth organ – hence the alternative name of labyrinthine fishes) is housed in the top of the gill chamber on each side of the head. The organs are hollow and are formed from the blood-vessel-rich skin lining the gill chamber. As the fish grows, the organs become more convoluted in order to increase the area of skin available for respiration. All anabantoids depend upon a supply of atmospheric oxygen for survival and some species will quickly die, drowned one might say, if prevented from reaching the surface of the water. A corollary of this is that some species can survive for a long time out of water, provided that they are kept damp. Many species of anabantoids construct floating bubble nests for the protection of their eggs and newly-hatched young.

Apart from the gouramis (strictly speaking the name should only be applied to the food-fish *Osphronemus goramy*) well known to aquarists, the anabantoids include the Siamese Fighting fish (plate 21), the Paradise fish and the Climbing perch. *Osphronemus goramy* is a large fish, growing to some two feet long. It is not particularly closely related to *Trichogaster*, but because they both have one or more of the pelvic rays elongated into a long filament, the same common name has been used for all. Explaining the origins of common names is often difficult and sometimes impossible: the so-called Kissing gourami (*Helostoma temnicki*) does not have any elongated rays in the pelvic fin.

One of the best-known anabantoids is the Climbing perch (*Anabas testudinosus*) of South-East Asia, whose ability to move overland and apparently climb trees, has been recorded in literature since 1797. The story originated from the observations of Lieutenant Daldorf, a Dane working for the Dutch East India Company in the then Dutch-owned region of Tranquebar. In 1791, during a heavy rainstorm, he noticed a fish in a palm tree about five feet above the ground. The fish was in a little stream of water in a crack in the palm's trunk where the narrow end of a broad frond channelled the rain. The palm tree was close to a swamp, whence it was assumed the fish had come. Further reports and local legends intimated that *Anabas* climbed trees to suck their juice. In some of Indian and Malayan tongues, the local names for *Anabas* mean climbing fish (in Thai, however, the local name is pla mor which means doctor fish).

In 1797, Daldorf published an article in the Transactions of the Linnean Society of London describing the extraordinary behaviour of this fish. As time passed, a certain amount of scepticism developed, especially among the Indian scientists who knew well the ability of *Anabas* to move overland but doubted that it could climb trees. The legend was eventually dissolved by Dr Das, who discovered that during its ambulations from one pond to another, *Anabas* was picked up by fish-eating birds and placed in the tree for later con-

sumption. Although we now know that it does not climb trees, the mechanics of its walking and its adaptations to a temporary terrestrial existence are of great interest.

The Climbing perch is less compressed and less deep-bodied than the Lace gourami. The body is covered with tough scales, the exposed edges of which bear spines. The thick skin and the spiny scales both act to reduce the rate of desiccation and to deter predators. The dorsal and anal fins have spiny rays and backward-pointing spines are found at the back of the sides of the head. As is mentioned above, the Climbing perch, because it is an anabantoid, can use atmospheric oxygen which helps to prolong its overland journeys. It seems that a major reason for its terrestrial excursions is to find other bodies of water which offer better living conditions, most likely more food. It has not reliably been observed to eat whilst out of the water. Most of its travelling is done at night when not only is the fish less liable to predation but the air is cooler and more moist, so the evaporation rate is lessened. As would be expected of such a fish, the coloration is subdued, a uniform dark brown sometimes with a few darker markings.

When on land, *Anabas* holds itself upright by spreading its pectoral fins to act as props. The body is then moved from side to side and the friction of the tail on the ground or against plant stems results in the fish moving forwards. The process is analogous to that of the fish swimming in water and the same movements of the body muscles are involved in both types of propulsion. It is not certain whether the pectoral fins directly contribute to the forward motion, by being used as "arms", or whether they merely act as props. The motion is very jerky, and has even been described as ungainly. Nonetheless a persistent fish has been reported to travel three hundred feet in half an hour.

Anabas is a food fish, and a particularly valuable one in hot climates, for its ability to stay alive out of water means that it can be carried long distances and will arrive fresh – an important and valuable asset in protein-poor regions. There is, however, a minor drawback to carrying live *Anabas* around the country and that is that the container must be well sealed, otherwise the fish will wriggle out and walk away. Dr E. W. Gudger reported that *Anabas* had been known to cause the death of fishermen. It appears that in some parts of South-East Asia, the fishermen will kill *Anabas* by biting off the head. Cases have been known of this action leading to the death of the man rather than the fish, the fish has twitched without warning, slipped down the throat and choked the fisherman to death. There the fish is lodged fast and, because of its backward pointing spines, removal is impossible.

For the benefit of aquarists, it should be stated that there are no records of aquarists having been killed by their gouramis. However, one does not suppose that aquarists attempt to kill their gouramis by biting off their heads.

Plate 22

A group of transparent fish. The Glass fishes, Chanda ranga *below, and* Chanda wolffi *above, and the Glass catfish,* Kryptopterus bicirrhus, *left.*

These three fishes have one outstanding feature in common – all are transparent. They are not the only transparent fishes, but they are the species most commonly kept by aquarists. The two fishes on the right are members of the genus *Chanda*, often called Glass fish, *Chanda ranga* and *Chanda wolffi*. There are over ten species ascribed to this genus and most of the smaller species are to varying degrees, transparent. The genus *Chanda* belongs to the family Centropomidae which includes, amongst others the giant Nile perch, a fish from larger rivers and lakes in Africa, which can grow to over six feet long.

The genus *Chanda* comprises marine, brackish and freshwater species living in South-East Asia and the Malaysia peninsula. In some cases, the same species will inhabit waters of varying degrees of salinity i.e. it is salt tolerant. Others e.g. *Chanda kopsii* are marine, while *Chanda baculis* is almost entirely a freshwater species.

Chanda ranga, the most frequently encountered species of the genus, comes from India. It was formerly known as *Ambassis lala*, but it was subsequently discovered that the fish so described was in fact the young of the previously-known *Chanda ranga*, hence the name change. *Chanda ranga* was (and possibly still is) common in the Ganges estuary and large shoals were caught so easily that they were used as fertiliser.

Chanda ranga is so transparent that the bones can be clearly seen, and, if the fish keeps still long enough, the heart can be seen beating in front of the silvery peritoneum. The peritoneum is the lining of the abdominal cavity hence most of the digestive organs, such as the liver and intestines are concealed. The swimbladder is clearly visible above the peritoneum.

The Glass catfish, *Kryptopterus bicirrhus*, is also from South-East Asia. As with the Glass fish, there are over ten species in the genus, but usually only one is imported. The genus *Kryptopterus* comprises pelagic catfishes with a long anal fin, a dorsal fin that is rudimentary or absent and two pairs of barbels. The majority of species grow to about five inches long, but one species, *Kryptopterus bleekeri* from Indo-China and Thailand, reaches two feet in length. *Kryptopterus bleekeri* is not transparent; its usual colour is pinkish or reddish.

The bones of the body of the Glass catfish are clearly visible, but again there is a silvery peritoneum lining the abdominal cavity. The swimbladder is conspicuous and looks like a bubble above the peritoneum. In one part of Thailand, the native name for this species is pla krayok which means windowpane fish and refers to the translucency of the swimbladder.

Transparency in fishes has evolved independently in several groups and in varying degrees. The species here are arguably candidates for the title of the world's most transparent fish. Africa has transparent catfish – one is *Physaila pellucida* from the rivers of western Central Africa. The body shape is similar to that of *Kryptopterus*, compressed and long. Its habits are also like those of *Kryptopterus*, it is a midwater species and as such unlike those of most catfishes. There is a transparent goby, *Aphya minuta* from the American Pacific Coast. Small specimens of some species of freshwater herrings from African rivers are translucent, if not transparent, the flesh is slightly opaque but the vertebral column can be dimly discerned.

Transparency cannot be equated with the absence of pigment. Obviously if pigment is present, then the fish is not transparent but the corollary that fish lacking pigment are transparent is not true. Some deep-sea fish are pure white, but not transparent; albino fishes and some cave-dwelling species lack pigment (and the silvery guanine layer is included under the heading of pigment in this context) but they are not transparent – just white.

Many larval fish are translucent or even transparent; there are several factors contributing to this condition, though. Firstly, the body is thin, i.e. there is less tissue to impede the passage of light. Secondly, the pigment layers have not developed. Even so, the degree of translucency varies from the larvae of one species to another. Although larval fishes usually lack scales, scales do not affect transparency, they are as subject to being transparent as is the body tissue. Scales are thin plates, which in most fishes are translucent (their very thinness predicates translucency) but not transparent. *Chanda ranga* has between sixty and seventy scales in the lateral line series all not just translucent, but transparent. The Glass catfish, like all catfishes, lacks scales. Naturally, all transparent fishes have thin bodies, a great thickness of any transparent substance is less transparent than a thin layer.

If it is not a contradiction in words, transparency can be coloured. *Chanda wolffi*, for example, although transparent, not infrequently has an amber or brown colour which scarcely renders it more opaque.

Transparency seems not to be a continuation of the larval condition or, if it is, then it is much enhanced. The opacity of every muscle cell must be reduced to the level of that of water. The translucent scales must become transparent and the pigment layers in the skin must disappear. Obviously there must be some survival advantage in being transparent; but the silvery peritoneum would seem to militate against the fishes' ability to disappear against the background.

We do not know the nature of the changes in the fishes' biochemistry that has allowed the largely undispersed passage of light through the tissue. We do not know if the mechanism is the same in the unrelated groups that have independently evolved this phenomenon. It is a process that is only maintained while the fish is alive for when the fish dies, the transparency is lost, much as H. G. Wells' invisible man became visible after death.

Plate 23

The Archer fish (Toxotes jaculatrix).

It has already been observed that the common names given to fishes are often inappropriate. This is true of the Archer fish (*Toxotes jaculatrix*) which would be better called "spitting" fish. The genus *Toxotes* comprises probably six species, all of which spit droplets of water to catch their prey. They live in both fresh and salt water from India and Malaysia to the northern tip of Australia.

Archer fish were first brought to the attention of the western world in 1763 in a communication to the Royal Society of London. On March 15th, Peter Ellison FRS read to the Royal Society the contents of a letter from Dr J. A. Schlober of Amsterdam:

> "Governor Hommel (Governor of a hospital in Batavia) gives the following account of the jaculator or shooting fish, a name alluding to its nature. It frequents the shores and sides of the sea and rivers, in search of food. When it spies a fly sitting on plants that grow in shallow water, it swims on to the distance of four, five or six feet, and then, with surprising dexterity, it ejects out of its tubular-mouth a small drop of water which never fails in striking the fly into the sea, where it becomes its prey."

The governor was so intrigued that he captured some Archer fish and put them into a jar filled with sea water, stuck a fly on the end of a thin stick and held it over the jar. "It was with inexpressible delight that he daily saw these fish exercising their skill in shooting at the fly with an amazing velocity and never missed their mark."

Hommel sent a specimen of the fish to the Royal Society, thereby causing some confusion, for the fish that he actually sent was the long-snouted Coral reef fish, *Chelmo rostratus*.

Further details of the behaviour of the Archer fish were given by Schlosser in 1766, and the German scientist Pallas published a description of it with a poor, but recognisable illustration.

For nearly one hundred and fifty years, no further details of the Archer fish came to light; rather, there was a strong suspicion among the zoologists studying oriental fishes that the reputation of the Archer fish was unjustified. Pieter Bleeker, an ichthyologist who had lived in Batavia, the same town as Hommel, wrote in 1875 that he believed its reputation was based on an error in observation. Dr Francis Day, an English surgeon and ichthyologist who wrote a monumental work on the fishes of India, denied the Archer fish its shooting power and instead attributed the ability to the *Chelmo* mentioned previously. This is understandable, because if the dead body of *Toxotes* and of *Chelmo* are compared, the long "gun-like" snout of the latter would seem to be the instrument more likely to be used, like a water pistol. Dr H. M. Smith in his 'Fishes of Thailand' points out that further confusion between the two species could have arisen because in Malay both fishes are called "sumpit-sumpit" which comes from "sumpitan" – a blowpipe.

It was not until 1902 that a Russian zoologist, Zolotnisky, kept *Toxotes* in captivity and confirmed that Hommel had been right. Zolotnisky also provided some other important observations. He stated that the fishes' aerial vision was acute and they could see small insects a long way off. He also noticed that *Toxotes* had very mobile eyes which "sparkle with seeming intelligence".

It has been suggested that nineteenth-century scientists were reluctant to believe in the spitting power of the Archer fish because they could not find a mechanism by which this could be achieved. The mystery was finally solved by Hugh Smith in 1936. He found that if he held a fish in the normal "firing" position and rapidly pressed its gill covers with his fingers he could, with practice, send a drop of water a distance of three feet. Detailed anatomical investigation followed until the entire system was understood.

The Archer fish has a long narrow groove or slot in the roof of its mouth. The tongue is thin and free at the front of the mouth but at the back it is thick and muscular with a median fleshy protuberance. When the tongue, with its median ridge, is pressed against the roof of the mouth it fits into the slot, converting the slot into a very narrow tube. The thin, free end of the tongue acts as a valve. When, therefore, *Toxotes* sees a likely fly, the tongue is pressed against the roof of the mouth, the gill covers are closed rapidly and the tip of the tongue is flicked to allow the drops of water to be forced out.

There still remains the problem of parallax. A stick poked into water seems to bend at the water surface and anyone who has tried to stand on a bank and spear an object underwater will have realised just how different are the refractive indices of water and air. *Toxotes* has the reverse problem. The keen eyesight and mobile eyes will certainly help, but recent research has shown that, for preference, *Toxotes* will try and get vertically below the prey. Incidentally, if it succeeds in getting just a few inches below a prey it will leap out and grab it rather than shooting it down. Archer fish normally cruise just below the surface and, as they frequently live in muddy water, the aerial vision is a great advantage.

Baby *Toxotes* can only shoot a couple of inches, and even then their shots are inaccurate. A fully-grown fish can be accurate at five feet from its prey. If the first shot misses, the aim is corrected for the second shot. For a smoker, *Toxotes* can be a nuisance. Dr H. Smith mentions that on two occasions a friend of his was smoking a cigarette near the water's edge when an Archer fish swam up and extinguished it.

Plate 24

A group of scats (Scatophagus argus).

In Greek mythology Argus, son of Inachus (or of Agenor), possessed eyes all over his head and body. Because of this he was known as "all seeing" and was persuaded by Hera to watch over the cow into which Io had been metamorphosed. Hermes decided to kill Argus and, according to one version of the story, Hermes played a flute until Argus was asleep and then cut his head off (another version has Hermes starving Argus to death). After the death of Argus, Hera transferred his eyes to the tail of the peacock.

Although one is not necessarily reminded of this story every time a peacock displays its tail, the memory is jogged by *Scatophagus argus*, the scientific name of the fish which is the subject of this painting. The spots on the body allude to the eyes of Argus. Sadly, the generic name is far less romantic and fanciful; scat derives from the genitive of the Greek "skor" meaning dung or faeces and phagus from the Greek verb "phago", meaning to eat. Were we to follow the Victorian naturalists' custom in coining the common name of a fish from the translation of the scientific name, the Scat would be named the thousand-eyed dung eater! In this description scat will be used as the common name.

Scats are widespread, they range from the east coast of Africa to the shores of the Malay Peninsula and from the coasts of India to Australia and Tahiti. They have been found along the shores of many of the Pacific islands within that compass, as well as in the estuaries and rivers of countries bordering the Indo-Pacific region. From the distribution it can be gathered that scats are euryhaline i.e. at home in fresh, brackish or salt water.

Although commonly the colour pattern is much as shown in the painting, throughout the area of distribution the colour patterns differ. For example *Scatophagus argus* from the east coast of Africa and the Malagache Republic lacks the spots but has instead about half a dozen vertical brown bands. Fish from these areas were at one time considered to be a different species. Even within one area the dark markings and the background colour can vary, as well as changing irregularly from juvenile to adult condition.

The larvae of *Scatophagus* are very different from the adults and undergo a gradual metamorphosis so that by the time they are approximately three-quarters of an inch long they resemble the adults. Scats (family Scatophagidae), and their close relatives the marine Butterfly fish (family Chaetodontidae), pass through what is called the "Tholichthys" stage. The use of this name for the larvae refers to the fact that Albert Günther originally described a larva as a member of a separate genus *Tholichthys*. (This genus is no longer valid, since the adults had already been named, but the name lives on.) The *Tholichthys* stages have a compressed, rounded body with a very large head and a tiny mouth. The head is covered above with conspicuous arched bones. A large bony plate lies in the skin in front of, and above, the pectoral fin. A bone above the gill cover carries a large backward-pointing spine and, of all the larval characters, this one is the last to disappear. As the young fish grows, the head becomes relatively smaller and the mouth larger. The bones in the skin become resorbed and eventually, when the fish looks much like the adult, the spine disappears. The advantage of these larval characters to the fish is not known.

The greater majority of fishes live in either fresh water or salt water, i.e. they are not tolerant of changes in salt concentration. The blood and body fluids of fresh water fishes are more concentrated than the fresh water, in which they live, but the reverse is true of saltwater fishes. Therefore in fresh water the body of the fish absorbs water. This phenomenon of water and salts flowing in opposite directions across semi-permeable membranes (such as the gills), in an attempt to equalise the concentration of salts on both sides of the membrane is called "osmosis". In freshwater fish, the kidneys are extremely important because they excrete the excess water that has permeated the semi-permeable membranes (i.e. diluting the blood), and also restrict the salt flow out of the fish. In a sea fish the problem is dehydration, for the water tends to leave the fish and go into the sea. The fish largely overcomes this by drinking lots of seawater and special salt-excreting cells in the kidneys eliminate the excess salts.

There are, however, some fishes like the scat (other examples include the Glass fish, the stickleback and the flounder) which can adjust very rapidly to waters of different salinities. The point to emphasise here is the rapidity of the adjustment which is in marked contrast to that of migratory species, such as the eel and the salmon, who spend a long time adjusting their physiology and for whom the reverse change is often impossible. The eel and the salmon have to acclimatise themselves to different salinities encountered during their breeding migrations; the eel only does it once and the salmon rarely more than once. Exactly how the scats can adjust their salt regulation so rapidly is not known. What we do know is that they can, with facility, cross barriers lethal to over ninety per cent of all other bony fishes.

To return to the generic name *Scatophagus*. In many parts of their range, scats have been reported as being attracted by and eating offal or sewage in the water. It is quite likely that they do this and certainly in many parts of the northern coastline of the Indian Ocean the fish is not eaten in the belief that it is unclean. In Australia, on the other hand, *Scatophagus argus* is regarded as a delicacy and is called the Spotted butterfish because of the softness of its flesh.

Plate 25